Luminos is the Open Access monograph publishing program from UC Press. Luminos provides a framework for preserving and reinvigorating monograph publishing for the future and increases the reach and visibility of important scholarly work. Titles published in the UC Press Luminos model are published with the same high standards for selection, peer review, production, and marketing as those in our traditional program. www.luminosoa.org

Virtuous Waters

Virtuous Waters

Mineral Springs, Bathing, and Infrastructure in Mexico

———

Casey Walsh

UNIVERSITY OF CALIFORNIA PRESS

University of California Press, one of the most distinguished university presses in the United States, enriches lives around the world by advancing scholarship in the humanities, social sciences, and natural sciences. Its activities are supported by the UC Press Foundation and by philanthropic contributions from individuals and institutions. For more information, visit www.ucpress.edu.

University of California Press
Oakland, California

Suggested citation: Walsh, C. *Virtuous Waters: Mineral Springs, Bathing, and Infrastructure in Mexico*. Oakland: University of California Press, 2018. doi: https://doi.org/10.1525/luminos.48

Library of Congress Cataloging-in-Publication Data

Names: Walsh, Casey, author.
Title: Virtuous waters : mineral springs, bathing, and infrastructure in Mexico / Casey Walsh.
Description: Oakland, California : University of California Press, [2018] | Includes bibliographical references and index.
Identifiers: LCCN 2017054276| ISBN 9780520291737 (pbk.) | ISBN 9780520965393 (e-edition)
Subjects: LCSH: Mineral waters—Mexico—Mexico City. | Baths—Mexico—Mexico City—History. | Groundwater—Mexico—Mexico City. | Water—Mexico—Mexico City—History. | Water resources development—Mexico—Mexico City—21st century. | Water-supply—Mexico—Mexico City—21st century.
Classification: LCC RM674 .W35 2018 | DDC 615.8/53—dc23
LC record available at https://lccn.loc.gov/2017054276

For Naomi, my little fish

TABLE OF CONTENTS

ILLUSTRATIONS

MAP

FIGURES

ix

I spent my childhood summers in a swimming pool in my best friend's backyard in suburban Los Angeles. Usually it was just the two of us and we would spend hours going down the slide, inventing new kinds of dives, or perfecting our splashing techniques. After lunch and maybe a monster movie on TV, we would head back for another few hours in the water. My family would also spend a week at beach at the end of every summer, and my sister and I would dedicate much of that to boogie boarding. I would fall asleep each night exhausted, feeling the swell of the waves as if I were still floating in the ocean. Southern California is dry, and maybe it was because the encounters I had with rivers and lakes were few and far between that the cold, clear freshwaters of the mountains filled my body and mind with a special kind of exuberant energy when I got a chance to swim in them. As I grew older, backpacking trips to the mountains became more frequent, and I had my first contacts with the strange, chthonic, social waters of hot springs. Throughout my college years and during a brief career as an archaeologist I was outdoors a lot, swimming and soaking when the opportunity arose. I learned to love water, and associate it with fun, freedom, adventure, and the beauty of nature.

Graduate school in New York City was followed by a long period in Mexico City, during which I conducted PhD research and then worked at the Universidad Iberoamericana. The dissertation I wrote during that period, later a book, was about the political economy and culture of irrigated cotton agriculture in northern Mexico, which provided me with a different, more cerebral engagement with water. The topic was fascinating, and I loved piecing the story together, but the water in that research never felt like all those waters I felt earlier in my life: hot, cool, surging with the waves, salty, invigorating, quenching, sulfurous. The water I read and wrote about wasn't really felt at all. It was a resource measured in cubic meters and liters per second; it belonged to cities, states, and nations and was the subject of treaties; it formed political boundaries, was delivered to farmers, moved by capillary action and evapotranspiration, was treated to comply with health standards.

After my daughters were born I began a slow return, something like a migratory fish, to the waters I knew during my youth. I moved from Mexico City back to Southern California, this time to UC Santa Barbara and its nearby beaches. Watching the faces of my kids as I bathed them as infants, or as I held them when they learned to swim in the pool and Jacuzzi at our apartment building, or jumped around with them in the waves at the beach, revived those registers of the human experience of water that I had forgotten, ironically, while I was becoming an expert on the topic. The somewhat utopian project of raising children pushed me toward less intellectual, more affective engagements with the environment and with other people, and I found myself back in the waters, playing, splashing, socializing. And thinking: How are these feelings and ideas about water shaped socially, culturally, historically?

This book is about Mexico, a country I love, my other home. And without all the people in Mexico the book would never have happened. I married Emiko Saldivar in Tlayacapan, Morelos, and our girls, Amaya and Naomi, were born in Mexico City. These three amazing women make me a better person, and give me all the reasons I need to work for a better world. The support and love of my extended family in Mexico has never wavered, and I gain strength from them all. I learned from Michiko about *onsen*, *ofuro*, and the importance of the bath; Américo shared with me a love of water that comes from growing up in a dry place. Aida Maria, Laura, Claudia y Serguei, Atis y Dani, all the Tios, Tias, Primos y Primas: I hope I haven't bored you with my stories of hot springs. Great friends in Mexico City always received me when I was turning pages in archives and libraries. Thanks to Margarita, Claudio and Ibó, Roger and Cris, Adriana and Bensi, Jorge and Catalina for the love and memories.

Colleagues have generously engaged me across all those institutional, intellectual, linguistic, and cultural borders, never questioning too much how my interest in the economic history of irrigated agriculture morphed into a project about swimming and bathing. Special thanks to Mario Cerutti, Eva Rivas, Arturo Carrillo, Cirila Quintero, Araceli Almaraz, and everyone else in the Asociación de Historia Económica del Norte de México, for all the intellectual engagements, comments on papers, and camaraderie over the years. And to Luis Aboites, a special thanks for providing detailed comments, yet again, on a manuscript that shrinks before the example of his scholarship. My deepest intellectual debts and bonds are with these colleagues, and I hope that with this book I have added something worthwhile to their excellent research.

Various institutions supported the archival and field research presented in these pages. The University of California, Santa Barbara provided travel and research funds through the Dean's office of the Division of Social Sciences and the Academic Senate. The Universidad Iberoamericana generously funded a semester of sabbatical residency, and provided access to the library and archive. A number of institutions invited me to present pieces of this book: El Colegio de la Frontera Norte; El Centro de Investigaciones y Estudios Superiores en Antropología Social; El Instituto Tecnológico de Monterrey; La Universidad Autónoma de Baja California; La Universidad Autónoma de Sinaloa; La Universidad de Extramadura; the Lancaster Environment Center of Lancaster University; the University of California, Merced; the University of Luxemburg. I thank you all deeply.

Lastly, I would like to express my deep appreciation to all those who work toward the goal of free and open intellectual production and publication for the benefit of humanity. The University of California is a public institution working for the people of California, and many here at UC take this mission very seriously. I particularly wish to thank the excellent editors at the University of California Press who have dedicated their time and energy to developing the Luminos Open Access Series, which offers knowledge freely to a public that transcends paywalls and border walls. The external manuscript reviewers and the editorial committee of the UC Press gave their time and energy to shaping the book, and the University Library and the Academic Senate at UC Santa Barbara provided the subvention that pushed this book into the realm of the creative commons. It has been my great pleasure to work with you all.

MAP 1. Mexico, with locations of hot springs discussed in this book.

Waters/Cultures

There were a few pages about Peñón de los Baños on the internet, and my guide-book also briefly mentioned it. I had thought it would be more important, con-sidering the presence of Peñón in the historical documents I was collecting in the archives downtown. Real hot springs in the middle of Mexico City—nature was difficult to locate amidst the densest of urban conglomerations. And because the shower seemed to have displaced the bathtub in my rented apartment, I was in dire need of a good soak. The bathhouse was located on Circuito Interior, the city's main circumferential artery, on a hill next to the airport, and occupied the lower floor of a nondescript U-shaped brick and concrete apartment building. In the courtyard garden of the building, however, a seventeenth-century chapel gave mute testimony to the powerful spiritual connections with these waters that once bubbled up from the earth on their own when this extinct volcanic hill was still an island in the lake that covered the Valley of Mexico.

As I looked around the place, I strained for glimpses of the uses, meanings, and practices sedimented in this site over time. In the foyer of the building a man ped-dling a spiritual cure invited me for a free diagnostic; there were flyers posted for energy alignment and a "course on miracles," as well as more common therapeutic treatments such as massage. Sitting in the drab hallway with a number of elderly patients and their attendants, I drank a few swallows of mineral water from a dis-posable paper cone and perused the old maps and photographs on the walls that testified to the prominence the place had once enjoyed as a sumptuous bathhouse and the site of bottling plants beginning in the 1880s, and then its renovation as a public health facility in the 1950s.

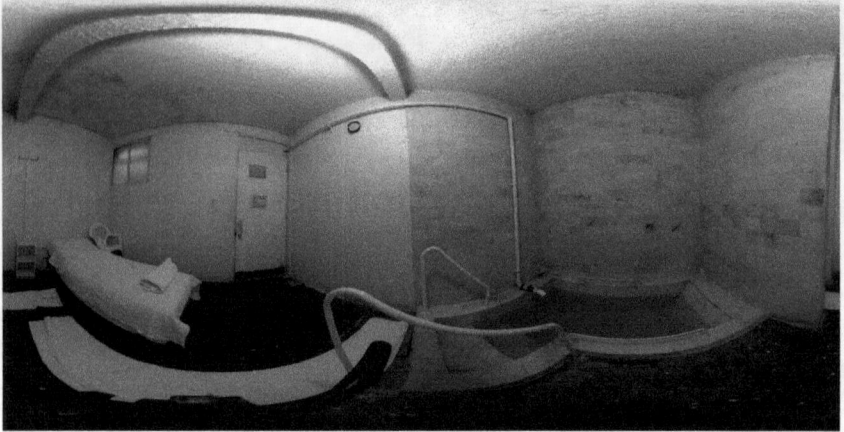

FIGURE 1. Individual bathing room with *placer*, Peñón de los Baños, 2017. Photograph by Author.

I was escorted into my own bath cubicle by a young woman in hospital scrubs and high, white rubber boots, and given precise instructions: soak for a maximum of twenty minutes; repose sweating on the cot wrapped in a sheet; do not drink more than three cups of water. The constant deep rumble of trucks and cars from the highway outside the window greeted me in the bathing room, where a tub of chipped, stained marble slabs—what was called a *placer* in Mexico for hundreds of years—filled quickly with steaming mineral water pumped from eighty meters below the building, a water that had been used for bathing in that locale for the last five hundred years. The sink did not work; the sheet covering the flimsy chromed cot with torn vinyl cushions was bleachy-clean but bedraggled. None of that mattered too much, because like the other clients of Peñón I was not looking for a luxury spa pampering. It was all about the water: soft and hot, relaxing and curative. A moment of natural healing in one of the world's biggest, densest cities.

HOT SPRINGS AND BATHING: EVERYDAY WATER CULTURES

The visit to Peñón's mineral springs in Mexico City was one of dozens I have made over the years in different places around the world. I of course am not alone: many people seek out mineral waters for their therapeutic properties, their flavor, their enjoyable warmth, and the good times they have drinking or soaking in them with friends and family. What began in college in the late 1980s as an encounter with the peculiar splendor of these strange corners of the landscape became twenty years later an anthropological research project to understand how and why people use these peculiar groundwaters in Mexico; how they imbue them with such intense and complex meanings; and how they understand their quotidian interactions with them, bathing and drinking. How is this most direct of experiences—drinking or soaking in water—a social and cultural construction? Who owns the mineral waters, and who is allowed access to them? What do the waters do to us? Why have scholars not embraced the topics of hot springs and bathing, when water is such a fashionable topic?

In this book I approach these as historical, anthropological questions, asking how interactions with mineral waters in Mexico took shape over the long modern period from around 1500 to the present. Most of what we know about the history of hot springs and bathing comes from Europe, and Mexico was conquered and colonized by Europeans, and so it is reasonable to start this inquiry there. While there is relatively little academic interest in mineral springs today, for millennia the study of their waters guided inquiry in Europe about the relation of humans to nature. Waters were thought of as multiple, unique liquids, much in the way that today we think of different bottles of wine as unique yet belonging to a unified category. Both everyday people and scientists also considered them efficacious— waters were agents that acted upon the world and, in particular, on the bodies of humans. The Romans are the most important influence in this quest to understand human-water relations; following the ideas of Greeks they formulated key elements of a medicine of waters that has lasted to the present day.[1] As they expanded their empire they built bathhouses on hot springs and incorporated local religious traditions into Roman bathing culture. Baths were ubiquitous in the Roman period, and still important throughout northern Europe during early Christiandom. During the years from about 500 to 1000 bathing became less frequent and took on new forms, but mineral springs retained their strong significance for healing, often in the form of holy water.[2]

After 400 AD, Roman infrastructure fell into disuse and Roman water culture fell into disrepute, but Mediterranean Arabs carried forth those ideas and physical engagements with water during the medieval period. Between the tenth and twelfth centuries much of the classical knowledge about bathing and health was translated and conserved by scholars in the western Caliphate in Cordova (now Spain), from where it spread to Italy and France.[3] For example, Peter of Eboli's narrative poem from the 1220s, about the thermal baths at Pozzuoli in southern central Italy, offers an extended discussion of mineral waters using classical references. The poem is testimony to a deeply ingrained popular culture of bathing with prominent sociality and sexuality that may have receded slightly between 500 and 1000, but was once again flourishing in the 1200s.[4] In Spain and other areas of the Arab world bathing in bathhouses continued to be a daily event for many people in the Middle Ages, and although the steam bath replaced the immersion pool as the principal practice, the waters retained their spiritual agency. In the 1500s, the ascendancy of Christians in Spain led to a century-long hiatus in which bathing was largely abandoned. In general, however, between 500 and 1500, people continued to take to the waters, giving lie to commonly held ideas of a dismal, dirty, and depressed medieval Europe.

As the elite renaissance of bathing expanded in France and other northern European countries in the sixteenth and seventeenth centuries, new users with new knowledge about the specificity and agency of waters butted up against existing ones. Scholars reread the works of Pliny the Elder, Aristotle, Hippocrates, Galen, and others who discussed the therapeutic benefits of bathing and drinking, especially in hot and mineral waters, and bathhouses were rebuilt using classical architectural plans.[5] Part of the impetus behind the reemergence of therapeutic bathing literature was to control the stream of people taking the waters, and shift the basis of popular healing practices from empirical "trial and error" toward disciplined reason in the hands of doctors.[6] Differences in hot springs water cultures were delineated by fundamental social divides, such as bourgeoisie/nobility and peasant/elite, but Renaissance science incorporated elements of late medieval popular culture that derived, albeit remotely, from the practices and knowledge of the Greeks, Romans, and later Arabs.[7] Beginning in the 1600s, immersion and steam bathing were recast as secular practices increasingly explained as therapeutic and, later, hygienic. Bathing and the use of heterogeneous waters grew dramatically in the second half of the nineteenth century, hand in hand with the extension of urban infrastructures and new structures of feeling about nudity, odors, cleanliness, and the availability of water.[8]

Curative and hygienic practices evolved that included bathing, showering, drinking, and even inhaling mineral waters, and each of these took many forms. Bathing by immersion was for a long time the principal form of contact with waters, and different kinds of bathtubs and pools were designed for different kinds of baths: the full-body bathtub, the smaller tub for the sitting bath, and even tubs for the isolated treatment of limbs. Each particular mineral spring was thought to have powers that derived from the specific qualities of its waters, and people sought treatment of their ailments by choosing among those springs, and from among the doctors who set up practices at each. The conceptual framework for understanding these waters and treatments evolved as well, as emergent scientific disciplines provided new information about the character of the different waters. Modern chemistry and medicine, for example, were pioneered in the spas of Europe as part of a search for the causes of mineral-water cures.[9] Even today, the labels of mineral-water bottles often provide analyses of the chemical contents of the waters they contain, and spas with mineral waters prominently display this information.

In the middle of the nineteenth century, Pasteur and the advent of microbiology shifted attention to the organisms living in water, and the effects of ingesting them. At that time drinking also became a principal focus for the prevention of disease and for curing ailments with mineral waters, as it was increasingly believed that the minerals in water were not absorbed through the skin, and therefore needed to be introduced through the stomach. The inhalation of mineral waters in the form of steam or mist also gained prominence in the late nineteenth century, while curative and prophylactic bathing focused on the physical application of streams of water to the body—showers, jets—which were also considered important for cleansing the skin of microbiological vectors of disease, and on the effects of the temperature of waters on the body. Physical contact with waters changed as medicine, science, and technology evolved.

The modern reshaping of water cultures gained momentum with the growth of capitalism and a reworking of human-environment relations in the eighteenth and nineteenth centuries. The modern spa was born in that period, linked to the formation of a bourgeoisie who engaged in leisure activities previously restricted to the nobility.[10] Historians connect European spas to the professionalization of the medical industry and its relation to state power, a process that formed part of the wider reconstruction of human relationships to water involved in public health and the sanitary city.[11] Increasingly, mineral and hot springs were destinations for middle-class urbanites seeking relaxation and therapy, and this movement constituted a budding tourism industry.[12] These particular social uses of mineral springs spread to other places in the world through the assemblages of empire, especially in the nineteenth century. Mineral springs were important sites of recuperation for French colonial administrators, for example, and hot springs bathing became an important activity in Brazil in the nineteenth century.[13] The British and Hapsburg empires created global networks of spas that served colonists and tourists, and the influence of Japanese bathing traditions is seen in Europe after 1860.[14] The business of bathing and bottling drove the reshaping of cultural engagements with waters.

Water cultures were formed not only in hot springs, of course. Ordinary Europeans bathed in rivers and lakes, especially where it was warm. Parisians with some money bathed at swimming clubs in the Seine as early as the thirteenth century, and people heated water to wash themselves.[15] Wealthy people took baths at home in the Middle Ages, and these, like the public bathhouses in areas under Arab influence, utilized regular water heated for the purpose. Seaside resorts also became fashionable destinations in the late eighteenth and nineteenth centuries, and experts produced theories about the therapeutic effects of bathing in these other waters, as well as elaborate "bathing machines" that lowered the delicate and infirm into the beneficial liquid.[16] Of course, many people just swam in whatever waters were nearby, for fun or to cool off.[17]

The popularity of bathing in public bathhouses culminated in the late nineteenth century. After that, the growth of urban hydraulic infrastructure moved the bath into the domestic setting, ending the era of the public bathhouse in many of the cities of Europe and North America by the early twentieth century. Simultaneously, advances in bacteriology called into question theories about the curative properties of water, and pushed medicine away from water. After the famous discovery by J. T. Snow that London's cholera epidemics were linked to water, the liquid was increasingly seen as a health risk rather than a benefit. Bottled spring water was favored by those who could afford it, but by 1900 the cities in the developed world built infrastructures and established water quality standards that assured clean, safe tap water for large urban populations.[18] In places where public water infrastructures were slower in coming or incomplete, or where deeper cultures of social bathing reigned, such as Eastern Europe and Japan, public bathhouses lasted deeper into the twentieth century.[19] Hot springs resorts fell out of style in many places in Europe and North America after 1920, while remaining popular among the middle classes in Spain and Eastern Europe, where national health care systems supported the water cure through the mid-twentieth century. The restructuring of economies around the world in the last decades of the twentieth century has changed our relationship to mineral waters once again. The neoliberal downsizing of public health systems created an opportunity for capital to refashion many of the spas in Europe as luxury establishments for a smaller, wealthier clientele.[20]

WATER STUDIES: HOMOGENEITY AND HETEROGENEITY

Considering the long history of mineral springs, and their importance to the bottling industry today, it seems strange that most contemporary water scholars ignore them and focus instead on the infrastructure and social organization of systems that use surface waters. Anthropologists, in particular, have written a lot about public water systems, but very little about mineral waters and bathing. Why is this? It is true that hot springs and mineral springs are quite rare compared to other sources of water, and produce a very small volume of water. Despite this, hot springs are very notable features of the landscape and have been the object of intensive use and cultural activity for thousands of years. They also have held the attention of scholars and scientists from the Roman period until well into the twentieth century.[21] In fact, mineral waters seem to have fallen from scholarly view only recently.

A more likely reason than their rarity for the neglect of mineral waters in research today is that they do not fit easily into modern narratives of water as a single, uniform, inert element that can be managed by a unified infrastructure. Christopher Hamlin and Jamie Linton have argued that for most of history waters were understood as heterogeneous, with distinct origins, properties, and powers.[22] This shifted in the eighteenth century, when the prevalent idea of waters as multiple gave way to the idea that water is a single, essential element: Lavoisier's formulation of all waters as H_2O. It was a movement of thought in which chemists, biologists, and sanitarians identified both the uniformity of water and a seemingly infinite variety of dissolved and microscopic contents that made each water distinct. In the emerging science of water, the liquid was a homogeneous element and the obvious qualitative differences among waters that so long occupied the attention of healers, city planners, farmers, and everybody else were now attributed to the "impurities" carried by those waters: sodium, iron, sulfur, carbonates, microorganisms, etc. In this cultural shift, heterogeneous waters began to share space, in an uneasy balance, with a unified water.

This paradoxical balance enabled diverse hydrosocial processes to unfold. The conceptual unification of waters into water was accompanied by the development of a new "arithmetic" style of reasoning that facilitated the management of large quantities of the liquid through extensive physical infrastructures. While the impressive hydraulic works of Rome, for example, certainly required sophisticated engineering to move large volumes of liquid, they were built to preserve the plural identities and agencies of the various waters that the city drew from different sources.[23] However, the conceptual shift from waters to water that began in the eighteenth century implied that an infinite number of sources could be brought together by a physical infrastructure extending limitlessly through Cartesian space. In this vision a singular water was subject to a single standard of quality that set acceptable amounts of different biological and chemical impurities such as bacteria, arsenic, and so on. The culmination of this process was the monumental integration of waters and waterways in the western United States after World War II, and the plans for even grander, transcontinental hydraulic works: a fully plumbed landscape.[24] As modern hydraulic infrastructures expanded, an ever-smaller proportion of people got their waters directly from wells, rivers, and the like, and more saw the tap or the irrigation canal as the source of water. That social, conceptual, and infrastructural shift to "water" obscured many of our uses of and knowledge about heterogeneous "waters" such as mineral springs.

HYDRAULIC SOCIETY, IRRIGATION COMMUNITIES, WATER CULTURES

The literature on water that developed in the twentieth century, including much of the historical and anthropological work on the topic today, reflects this intellectual and infrastructural domination of "waters" by "water." Rather than devote energy to understanding particular waters and how they shape diverse human ecologies, scholarship on water in the twentieth century usually treated water as an inert, universal backdrop to the question of how humans organize themselves socially and politically to utilize the substance. These water studies can claim one origin in the work by V. Gordon Childe and Karl Wittfogel that theorized the connection between the rise of complex societies and state power and the physical and political control over water. Anthropologists Julian Steward and Angel Palerm incorporated Wittfogel's ideas into the "cultural ecology" perspective in anthropology that they developed in the United States and Mexico in the 1940s and 1950s.[25] Scholars working in this tradition centered attention on the control of water to produce agricultural surpluses, the constitution of peasant and political classes, and the transfer of surplus from the former to the latter.[26] For decades since, debates have wheeled around the central pivot of irrigated agriculture and the state, and histories of water that focus on the modern period in both capitalist and socialist systems reproduce the same assumptions about water as an undifferentiated and inert backdrop.[27] In all this work the water itself is assumed to be a homogeneous substance across diverse geographies and cultures, incapable of influencing people's bodies, their activities, or their ideas about the environment.

The critique of how irrigation served to consolidate the power of state and capital compelled many anthropologists studying water to turn their attention from "hydraulic society" to small-scale irrigation systems managed by peasant communities.[28] These scholars questioned a central assumption of the "hydraulic society" literature by pointing out that irrigation does not necessarily lead to despotism or state formation, but is often at the heart of the reproduction of community and peasant domestic economy.[29] Aspects of culture such as authority and religion were recognized as playing a key role in water management. But despite the critical angle taken by the "irrigation community" literature on the high modernist pretensions and failures of large-scale irrigation, it shared with the work on "hydraulic society" a common understanding of water as a uniform substance to be managed, and it privileged questions of social organization and technology. The differences among systems were found in the structure and scale of water management, more than the cultural understandings of the water itself, or the plethora of uses people make of waters in their daily lives other than irrigating fields and managing hydraulic infrastructure. The unitary, arithmetic notion of water as a singular, exchangeable substance persisted, carried forward in the culture of scholars and politicians who, despite the differences in their political projects, shared an ontological blindness to the heterogeneity and efficacy of waters.

Despite the rise of homogeneous water among scholars and planners, the heterogeneity of waters and water cultures never disappeared, and actually gained strength through the business of bathing and bottling. Even today people discern the particular characteristics of waters in different public water systems: New York has famously good tap water, Florida not so much. But it was mineral waters that retained their identities most strongly. During the eighteenth and nineteenth centuries, while other waters were physically integrated into infrastructural systems, mineral waters were left to themselves and their ancestral uses. Their dissolved minerals often render them harmful to agriculture fields, industrial machinery, and urban pipes, and so these heterogeneous waters continued to be used for bathing and drinking, activities that expanded during the eighteenth and nineteenth centuries, reaching a peak around 1900 before fading in the 1920s. Hot springs spas flourished all over Europe, the United States, the French and English colonies, and, as we shall see, Mexico. During the late nineteenth and early twentieth centuries, mineral waters and other watery drinks became a major industry, and today the business of bottling heterogeneous waters is expanding once again.[30]

The recent proliferation of heterogeneous watery use-values takes place in the context of a global "water crisis" defined by serious contamination problems and an absolute scarcity of the resource brought on by waste and hard limits to the amount of fresh water that can be captured and stored with infrastructures we have built over the last century. Water managers in the United States realize that the construction of yet more massive, elaborate, and energy-intensive hydraulic systems is not a sustainable solution.[31] The World Bank and other national governments have followed suit in seeking less costly infrastructural solutions to providing water for irrigation and urban use, and placing more emphasis on decentralized organizational and political solutions to reducing overall consumption of the liquid.[32] This turn to decentralized demand management has brought with it a recentralization of water management in the realm of culture.[33] But the concept of "culture" at work here is often narrow and instrumental: shared economic and environmental values for the liquid to be distributed from the top down. Where possible, demand management programs start by setting prices for water that will lower consumption. Usually these are tiered pricing schemes in which a basic quantity of the liquid is assured at low or no cost, and greater amounts can be purchased at increasingly higher unit rates. The assumption behind these schemes is that high prices serve as signals for consumers to reduce their consumption. "Culture," from this perspective, is a unified system of values, shared within a group, that guide the universal, economic decision-making of rational individuals.

Decentralization and demand management in the neoliberal moment have also been initiated from below as popular processes, and these movements are rooted in deep cultural histories and local meanings for waters and landscapes. For example, the "New Water Culture" (Nueva Cultura del Agua) movement in Spain came together in the 1990s to recover, foster, and create environmental ethics and participatory management.[34] Activists and scholars argued that irreplaceable elements of their environment, society, and culture were threatened by the government's 1992 National Hydraulic Plan, and they spearheaded an effort to chronicle and valorize the multiple uses, values, and meanings of the water.[35] Another example of how sensitivity to local meanings and waters is being propelled by local action comes from the Standing Rock Sioux Tribe Reservation, in what is today the U.S. states of North and South Dakota. Thousands of people from all walks of life have joined the struggle of the Lakota Sioux to defend their lands, their waters, and themselves from contamination and dispossession by oil companies and their government allies. A key phrase in this mobilization is "water is life," which expresses an unyielding respect and love for the planet and its beings that is at odds with a way of life built on extraction. The politics of water has clearly moved onto the terrain of culture.

THE POLITICAL ECOLOGY OF WATERS: NEW
MATERIALISMS, OLD ONTOLOGIES

Scholars are contributing to this upsurge in interest in water cultures in at least two ways. On the one hand, they are producing studies on a number of important topics, including the varied and complex meanings for water,[36] practices of swimming and bathing,[37] long-standing uses and meanings of mineral springs,[38] the current boom in bottled waters,[39] and role of science in shaping our interactions with the liquid.[40] Along with these new topics of study, scholars are exploring new ways of theorizing and depicting the relations humans have with the world that surrounds them. The modernist assumption of the centrality of human will, intentionality, and action has been roundly questioned, and a whole array of animate and inanimate nonhuman agents are now contemplated as participants in "assemblages" or systems that make history and act politically.[41]

The foundations for this scholarly perspective of "new materialism" are often found in Spinoza, Deleuze, and other philosophers, but in this book I suggest that nondualist ontologies of material vitality and efficacy permeate popular culture, and can be identified in the history of mineral waters and bathing. For thousands of years people have ingested and immersed themselves in mineral springs because they believe these waters have a beneficial effect on their bodies and souls. These waters are still considered to be efficacious, as evidenced by the immense market for bottled mineral waters and mineral water–based cosmetics. This is not simply the idea that pure waters do no harm and dirty waters are bad for you, but rather that mineral waters are "virtuous"—that they are powerful agents that act beneficially and therapeutically on the human organism to increase well-being. Drinking and bathing in mineral waters are activities motivated by a popular ontology not entirely commensurable with that which holds the individual human self to be sovereign.

To understand the long history of this popular ontology of waters this book takes a political ecology approach to the social relations and cultures of mineral waters, bathing, and infrastructures. Political ecology infuses a materialist focus on human-environment dynamics, social organization, and power with a critique of the conceptual categories that structure socioenvironmental inequality and destruction.[42] Political ecology thus urges us to consider how a modern ontology of water came to dominate other ways of understanding waters as a material, social process. The book shows how conceptual dimensions of the waters/water dynamic are connected to the expansion of hydraulic infrastructures, the integrated of waters and people into coordinated hydrosocial systems, the displacement of some forms of bathing by others, the inclusion of heterogeneous waters in commodity exchange, and the role of the mineral waters themselves in shaping all of this. But political ecology also helps us to recognize that this historical process of domination is neither unilineal nor complete, and that alternate concepts and uses of waters continue to exist together with the groups that nurture them.[43]

CHAPTERS AND ARGUMENTS

In 1500 Iberoamerican water cultures were marked by deep conflict. In chapter 2 I use secondary literature and firsthand accounts of soldiers and priests to discuss how, at the close of the Reconquista, ascendant Christians in Spain attacked Jewish and Arab institutions and practices of bathing, especially the *hammam,* or steambath, driving the bath out of sight in the sixteenth century. Conquering Spaniards in what is today central and southern Mexico brought this deep hostility toward bathing to bear on the indigenous steambath, or *temazcal,* which was the principal mode of bathing in the Americas and an important site for social, therapeutic, sexual, and religious activities. Bathing in water recovered its acceptability by 1600, although Spanish missionaries and government officials continued the effort to extirpate indigenous cultural practices from the *temazcal* and reduce its multiple functions to only the cleansing of bodies. By 1700 the *temazcal* was widely accepted among American-born Spaniards, and many indigenous people and humble mestizos also periodically immersed themselves in hot water bathtubs (*placeres*) offered by the bathhouses in Mexico City. During this period the first evidence appears that Mexican hot springs were developed into baths by religious orders to treat diseases and ailments, part of a burgeoning transatlantic field of medicine that carried with it the revaluation of hot and mineral spring-waters. Records of popular bathing for health and pleasure in the hot springs of Peñón de los Baños and Michoacán also appear at this time, showing that this was a cultural shift that worked its way throughout society. This chapter discusses the intersections of class and race that shaped bathing and the social use of hot springs in colonial Mexico, and shows how these cultures of water were shaped by hierarchical fields of power, notions of bodily difference, and inequality in access and property.

Chapters 3 and 4 focus on bath practices and water science in the Enlightenment. The late eighteenth century is a particularly important moment in which notions of cleanliness, public health, and urban order came together in the reorganization and regulation of Mexico City's water system and the practices and meanings of bathing. Chapter 3 shows how ideas of rational government were deployed to deal with problems of water scarcity and social effervescence. Investments in infrastructure brought together multiple waters, and the material and conceptual unification of waters as a singular substance began to take shape unevenly. Mexico City suffered from a scarcity of freshwater, and so the viceroy Conde de Revillagigedo launched a campaign to extend and improve the hydraulic infrastructure. These material developments were accompanied by a moral effort to reshape popular bathing practices that were deemed dangerous to boundaries of race, class, and sex, and the social order those boundaries defined. Archival documents from the Departments of Water and Police of the Mexico City government attest to efforts by the ruling class to circumscribe popular bathing practices and discipline unruly subjects. Much of this was aimed at keeping people and waters in the right place:

local officials intervened to stop people from bathing themselves and their animals in the public fountains, to keep men and women apart in the bathhouses, and to keep wastewater separate from freshwater. The arts of government were enacted in the spaces of the bath and on the bodies of bathers in a quest to form modern moral and political subjects.

The late eighteenth century also witnessed the emergence of science, and Mexican mineral waters were a principal object of study for chemists, pharmacists, and doctors. In chapter 4 I discuss how mineral springs medicine flourished during a time marked by intellectual and cultural opening and, eventually, the dismantling of the Spanish colonial government in the Americas. Studies of various Mexican hot springs were carried out under Royal orders in Michoacán, Tehuacán, and the Valley of Mexico, and the church conducted other studies. It was during the rule of the Bourbon government that the bathhouse at Peñón de los Baños was rebuilt, a sign of the prosperity generated by increased trade as well as technological advances in mining and industry. Growing wealth and the upwelling of scientific ideas about the efficacy of waters only partially displaced, however, everyday practices of bathing and access to these waters by the poor.

Chapter 5 shows how, in the second half of the nineteenth century, improved drilling and pumping technology integrated subterranean aquifers into urban water infrastructure, providing an unprecedented opulence of water. New sources of groundwater facilitated the creation of many new public bathing facilities in Mexico City and a related reduction of the flow of springs that served local communities in the Valley of Mexico for thousands of years. Swimming pools and bathhouses opened in the new, wealthy neighborhoods near Chapultepec Park and along Paseo de la Reforma, marking an explosion of social bathing. A period of exaggerated economic growth between 1890 and 1910 supported a dramatic expansion of the urban water system, the building of household bathrooms, and the practice of individual private bathing. This marked the beginning of a long, and never fully consummated, shift from public bathing to private bathing.

Chapter 6 shows how this changing sociality of bathing in the nineteenth century was accompanied by advances in chemistry, microbiology, and medicine. Journal articles from the mid-1800s tell us about the project to scientifically characterize the diversity of the waters used for bathing and drinking in Mexico. These documents reveal how microbiology defined established practices of bathing and drinking as potentially dangerous for public health, and set new parameters for the healthful interaction with water. Biological approaches did not displace chemistry from its position of authority in the realm of public health; in fact, belief in the therapeutic virtues of mineral waters only increased. Businesses of mineral water treatments—bathing and drinking—were established in Peñón de los Baños, Guadalupe, Topo Chico, Aguascalientes, Tehuacán, and elsewhere, and these businesses reinforced concepts of heterogeneous waters and alternate bathing practices.

The development of businesses at mineral springs did not occur in a vacuum. In chapter 7 I evaluate the role of the Mexican state in promoting the transfer of control over mineral springs from communities of peasants to urban industrial businessmen. The incursion of capital into Mexico during the late nineteenth and twentieth centuries led to the renovation of bathing facilities after almost a century of neglect, the creation of mineral water bottling plants, and the privatization of mineral and hot springs. In Tehuacán the state facilitated the consolidation of the bottling industry by imposing public health regulations that eliminated small and artisanal companies. State regulations insisted on a homogeneous standard of biological quality that enabled bottlers to increase their production of heterogeneous mineral waters and soft drinks. In Topo Chico, state lawyers and scientists helped to wrest control of the waters from peasants, who for centuries relied on them for agricultural and domestic uses, and place them in those of industrial bottlers including the Coca-Cola Company.

In chapter 8 I argue that the ongoing heterogeneity of water cultures is rooted in social heterogeneity. The pressure on hot springs generated by capital expansion into Mexico between roughly 1880 and 1930 met with strong resistance by rural, small-town Mexicans who fought to maintain their waters as common property with open access. After the revolution, national elites inspired by the model of tourist development put into practice at the Agua Caliente hot springs in Baja California and in Tehuacán collaborated with local actors in an effort to turn the town of Ixtapan de la Sal, in Mexico State, into a destination for bourgeois tourists from Mexico City. However, residents of that town challenged the new monopoly by outsiders over the hot springs they had always used, and charted an alternative plan for community ownership and management of those waters that preserved access to them for locals and humble visitors. I argue that this struggle and others over Mexico's mineral springs were brought on by competing cultural projects defined in terms of race, class, ethnicity. and locality.

I conclude on a positive note. There often seems to be little hope for restoring a respectful relationship with the waters in our world. The construction of massive infrastructure proceeds apace, and groundwater in all parts of the world is rapidly being depleted. Visit most households in most cities and the unification of water appears to have gotten the upper hand: people are hard pressed to identify their water, its qualities and origins, and most have just as little understanding of the infrastructures that serve to connect them to the world and each other. Sit a while in a hot mineral spring, however, and the people you meet will explain the particular qualities and therapeutic uses for that spring, and compare them with those of other springs: some salty; some sulfurous; some metallic. The springwaters you soak in will leave your skin feeling and smelling a certain way, which may compel you to consider how those particular waters act upon your body.

This book concludes that such experiences are vitally important to any project of reconstructing our relation to water, and that our daily interactions with water—bathing and drinking in particular—are potential sites for this reconstruction. Most efforts to deal with problems of scarcity and pollution of water try to increase supply or decrease demand through modern universalizing approaches such as monumental infrastructures or water markets. These approaches have not worked so far, and the book suggests that they may be part of the problem rather than the solution. Our modern water cultures are relatively recent developments, and even today not all aspects of our water cultures are alienated and homogenized; they never fully will be. Water cultures are products of long material and meaningful histories that we can trace back hundreds or, in some cases, thousands of years. And while the economic, social, and cultural dimensions of modern water may push us toward integration, uniformity and exchangeability at ever-greater scales, they never fail to reproduce heterogeneity. The wealth of varied practices, ideas, and values that make up this heterogeneity may help us to move our relationship with water in a more sustainable, less damaging direction.

A NOTE ON THE TEXT AND METHODS OF HISTORICAL ANTHROPOLOGY

This book uses techniques from history and anthropology to tell a long story about waters and people. It deploys mostly archival and documentary evidence to locate the origins and describe the evolution of our relationship with waters in Mexico. In this sense it is cultural and social history. However, I spent days, weeks. and months at hot springs in Mexico, bathing and socializing, taking interviews and notes, and drawing site maps. That fieldwork is not visible in the text, but it frames the historical research, and defines many of the questions I hope to have answered in the book. What may be more apparent is the ethnographic approach I take to the archival and documentary evidence, always looking for the quotidian experiences and cultural understandings of people who drank and bathed in the past. I reproduce, verbatim, the words and testimony of participants in this history, and set these passages apart in boxes, in quotation marks. In other places I reconstruct what I imagine was happening from the perspective of those participants. These reconstructed passages are also set apart in boxes, but have no quotation marks, as they are, finally, my own words.

2
———

Bathing and Domination in the Early Modern Atlantic World

"From this place we could likewise see the three causeways which led into
Mexico—that from Iztapalapan, by which we had entered the city four days ago;
that from Tlacupa, along which we took our flight eight months after, when we
were beaten out of the city by the new monarch Cuitlahuatzin; the third was that
of Tepeaquilla. We also observed the aqueduct which ran from Chapultepec, and
provided the whole town with sweet water. We could also distinctly see the bridges
across the openings, by which these causeways were intersected, and through
which the waters of the lake ebbed and flowed. The lake itself was crowded with
canoes, which were bringing provisions, manufactures, and other merchandise to
the city. From here we also discovered that the only communication of the houses
in this city, and of all the other towns built in the lake, was by means of drawbridges or canoes. In all these towns the beautiful white plastered temples rose
above the smaller ones, like so many towers and castles in our Spanish towns, and
this, it may be imagined, was a splendid sight."

—Bernal Díaz del Castillo

Source: Díaz del Castillo [1568] 1844.

The Spaniards were astonished. The city they entered was clean and orderly, with
enormous bustling markets, wide streets and plazas, huge pyramids, and other
impressive feats of engineering. In some ways their descriptions recalled the early
modern Iberian cities they knew, with white-plastered monumental architecture,
peasants and nobles, and thriving regional economies. Unsettling the comparisons

15

at a fundamental level, however, was the fact that Tenochtitlán was an aquatic city, built to float at the edge of land and water, different from those of the semi-arid landscapes back home. Díaz del Castillo's perspective on this city in the lake was gained from the top of the Templo Mayor—the main ceremonial pyramid—on November 12, 1519, and was recounted in Spain in 1568, a lifetime after the Spaniards conquered the Mexica rulers and subjected the people. Much of the wonder of the experience had faded by the time of the retelling, and what remained was a stra-tegic, military perspective that identified the important points of control over the watery milieu: drawbridges, aqueducts, and causeways. In the twenty-one months that followed the visit to Moctezuma's palace, these crucial infrastructures were destroyed by the Spanish and their allies. The domination of the lacustrine capital of the Mexica empire was not simply a military campaign, however, and it did not end in 1521. Over the centuries that followed a slow siege was laid on the underlying relationship between the waters of the Valley of Mexico and the human, built envi-ronment. The pre-Hispanic water culture—infrastructures, ideas, and practices—that formed as an adaptation to that place was drastically reshaped in the ongoing crucible of conquest.

In this chapter I trace the long process of change in the water cultures of Mesoamerica by focusing on struggles over bathing—over the direct, intimate, bodily contact between people and waters. I focus here on central, highland Mesoamerica, especially the Valley of Mexico, because it was the most densely populated area, with many hot and mineral springs and substantial historical docu-mentation. It is, however, only one region of what is today Mexico, and although much of this book is focused on this region, we shall see in later chapters that people in other places bathed and otherwise engaged with mineral springs and waters in different ways. Bathing is a topic of inquiry that has not been explored in Mexican history, partly because documentation of this aspect of water culture is scant, but also because Díaz del Castillo's gaze from the top of the Templo Mayor reveals a blindness shared with scholars—even environmental historians—to the most common, quotidian interactions with water. This absence is created by the overwhelming presence in the literature of more strategic questions of hydraulic infrastructure and state formation. The literature on water has lavished attention on irrigation and agriculture, but proportionally few people in the long history of Mexico irrigated anything. Bathing and washing, on the other hand, are the com-mon contacts with water that most people in the world have, including those who build, manage, and operate irrigation systems.

The verb "to bathe" signifies, at the very least, a contact between the body and water, or some other substance that behaves in a similar way, such as sun, dust, or light. When used to talk about water, the word can signify a range of different encounters with the liquid. Usually, in today's English, bathing is thought of as an act of cleaning one's body in water, or simply washing parts of one's body with water. Bathing can mean immersion in a tub of hot or cold water, using a wet cloth or sponge to clean the entire body or parts of it, standing under a shower, or even sitting in a room filled with steam. Immersion or other contact with water that does not involve soap or shampoo is less commonly conceived of as bathing, but actively swimming or diving through water, or simply lounging about in water, is sometimes also described as bathing, as the term "bathing suit" attests. In today's Spanish, these overlaps are similarly evident, as the verb to bathe (*bañar*) is frequently used to talk about immersion in the ocean or the swimming pool. The physical activity of swimming for exercise is now more often denoted by the verb "to swim" (*nadar*). Bathing in steam, a common practice throughout the Mediterranean world, Scandinavia, and Eastern Europe, was also the principal form of bath for the people of Mexico before and after conquest, all the way up to the nineteenth century. This Mesoamerican steambath—the *temazcal*—was called a *baño* or "bath" throughout the colonial and national periods, and the bathhouses, or *baños*, of Mexico would usually include tubs for immersion as well as a *temazcal*. At different moments in different places, these meanings and practices blurred even more than they do today.

TENOCHTITLÁN: THE CITY IN THE LAKE

"The whole body of the city is in the water."

—Francisco López de Gómara

Source: López de Gómara [1552] 1966: 147.

The Spaniards concentrated themselves and their activity in the highland plateau of what is today central Mexico, especially the lake-filled Valley of Mexico, home of the Aztecs. During much of the twentieth century Mexico City held the title of most populated metropolis in the world, stretching over nine hundred square miles in a valley surrounded by mountains. Apart from the rainy season between June and August, it is dry. There are some parks and open spaces, including wetlands and lakes near the airport to the east, and in the south in Xochimilco and Chalco. These watery zones, and the fact that parts of the city flood regularly during the rainy season, are reminders that this most urban of spaces was once a vast shallow lake fed by rivers coursing down the slopes that surround the city center. Place names also signal the hydraulic foundations of the city. Santa Maria la Ribera, for example, a neighborhood northwest of the downtown, was on the shore of the lake until the nineteenth century (*ribera* means "shore" or "bank"). Many roads were built on top of rivers that were turned into drainage tunnels as the city grew: Río Magdalena; Río Churubusco; Río de la Piedad. The waters themselves are hard to find now, as invisible as those tunnels that channel the huge volume of water falling as rain each summer northward out of the Valley of Mexico to the Río Tula. This enormous drainage project demands its opposite: an equally monumental system that brings a flood of freshwater to the city's pipes and faucets from hundreds of kilometers away and a kilometer downhill.

During the few centuries before Díaz del Castillo stood on top of the Templo Mayor, a water culture evolved in the Valley of Mexico that was fundamentally unlike that which we know today. The inhabitants adapted to living in a lake by building a highly productive agricultural system of raised-bed fields (*chinampas*) on the shores and shallows of the lake that enabled three harvests of the principal crops—maize, beans, and squash—by maintaining moist soil through the long dry season. Remnants of these fabled "floating gardens" can still be found operating in the southern end of Mexico City, in Xochimilco and Chalco. The plains and hills that surrounded the lake were dry-farmed during the rainy season, or supported with irrigation through the construction of dams and canals for surface waters and shallow wells for groundwater.[1] The *chinampas* were also used as nurseries to produce seedlings that were transplanted to fields farther from the lake once the rainy season commenced.[2] The lake itself—shallow, warm, and bathed in tropical sunshine—was enormously productive, providing all sorts of food and other useful materials, and hunting, fishing, and gathering these resources continued to supply much of the animal protein and other important nutrients up through the nineteenth century, as well as raw materials used to produce baskets and mats, the roofs of peasant houses, and many other household objects.[3]

The surplus generated by these activities supported population growth, urbanization, the constitution of political, warrior, artisan, and intellectual classes, and the creation of an empire.[4] The Aztecs formed out of an alliance between the Mexica who had settled on the island of Tenochtitlán (where the historic center of Mexico City is today) in 1325, and their less-powerful partners in Texcoco and Tlacopan. Together they could mobilize upward of a hundred thousand soldiers, and thus were able to defeat the lord of Azcapotzalco in 1428 and dominate the Valley of Mexico until the Spanish arrived in 1520. The Aztecs, and especially the rulers of Texcoco, were skilled hydraulic engineers who mobilized the same masses of subjects who fought as soldiers to build dikes and causeways with roads that complemented a network of shallow channels dug into the lakebed to facilitate canoe traffic.[5] The island-city of Tenochtitlán reached a population of eighty thousand people at its height, fed and supplied by the lake and by the subjects of its far-flung empire.

Water is unpredictable and powerful. As the city of Tenochtitlán grew, it responded to the destructive behaviors of the lake with increasingly sophisticated engineering works that did not so much seek to eliminate the water as tame it. In the 1440s floods ravaged the city, driving the rulers to take dramatic measures to protect it from further inundations. In response to the 1446 flood the ground level of the city center, with its ceremonial buildings, was raised by the city's residents about two meters, and in 1449 Nezahualcóyotl, the ruler of Texcoco who was allied with Moctezuma Ilhuicamina, the ruler of Tenochtitlán, designed and built an enormous earthen levee across the entire lake, protecting the city as well as the rich agricultural lands and the fresh waters of the western shore from the salty waters that surged into the eastern end of the lake.[6] This dike, known as the Albarrada de Nezahualcóyotl, was 10 feet high and almost 25 feet wide, and stretched from north to south for some 16 kilometers—an especially mind-boggling achievement considering there were no beasts of burden in Mesoamerica to do the heavy lifting. At the same time that these building techniques kept lakewater out of the city and fields, they also supplied Tenochtitlán with clean water. In 1426 the Mexica ordered the construction of a raised, two-channel aqueduct that crossed the lake from the Chapultepec springs.[7] They relied upon the expertise of Nezahualcóyotl and his fellow architects from Texcoco, who shortly before the Spanish arrived built an irrigation system in the eastern foothills that extended some twenty kilometers and bound five towns together with shared infrastructure and managerial institutions.[8] The enormous amount of social labor required for all these infrastructural works was commanded through compulsory tribute obligations, which led some scholars to consider the Aztec empire a form of "irrigation civilization" similar to those in ancient Egypt, China, and Mesopotamia, where water was controlled by a supremely powerful state.[9]

Our understanding of large-scale processes of infrastructure construction, agriculture, urbanization, and state formation in the Valley of Mexico before conquest is relatively solid; we know much less about the daily activities that formed the substance of those processes. People literally lived in and on the water. The buildings in some of the villages on the lakeshore, such as Coyoacan and Iztapalapa, were built on stilts so that the rising and receding lake waters could pass beneath them. The seasons were marked by the ebb and flow of the lake as it filled with rain and dried again, the life cycles of the flora and fauna that lived in the lake, and the livelihood practices that depended on that water. Hunting, fishing, and collecting provided much of the animal protein, often in the form of insects and their larva, as well as materials for houses and household objects.[10] Salt, a key necessity for the largely vegetable diet, was extracted from the salt flats on the eastern shores of the lake by washing the soil and boiling the resulting brine.

The commingling of land and water in the Valley of Mexico was mirrored in the ideas and beliefs of the people. The indigenous people in the central highlands of Mexico, and throughout Mesoamerica, shared a complicated understanding of the constitution and order of the universe, and the place of people in it. In this "cosmovision" the land was considered to be surrounded by water far to the east and west, and water filled the depths underneath the land.[11] The hills were permeated with water, and the springs and rivers that sprung forth from that watery land were met by the celestial waters of the rain. The god Tlaloc ruled over this watery underworld realm, as well as the lightning, thunder, and rain that were generated by the mountainous heights and fell from the skies above. Water was both the source of life and fertility as well as a worrisome and destructive force, whose complacence sometimes required the sacrifice of children. This cosmovision mapped onto the experience of living in the landscape of lakes, islands, wetlands, and canals of the Valley of Mexico.

The fields, towns, and cities of the Valley of Mexico were saturated with lakewater, and wading, swimming, and diving were daily activities.[12] There are few comments about these kinds of activities by Spanish or Indian chroniclers, however. During the conquest and early colonial period the Spanish observers noticed the cleanliness of the people and the cities, and the frequent washing and bathing of all ranks of people. The cities of Tenochtitlán and Texcoco built urban water systems for public use, and in the streets of Tenochtitlán there were public latrines. Chroniclers of the conquest of Tenochtitlán remarked upon the orderliness, amplitude, and particularly the cleanliness of the public spaces, where human waste was collected and transported to the agricultural fields so that agricultural production was increased and little sewage entered directly into the water.[13] The houses of the small noble class in Tenochtitlán were plumbed for water, and in Moctezuma's palace there was a "beautiful fountain with lots of water that flowed through underground pipes to other parts of the house."[14] The houses of the elite also featured private steambaths, or *temazcales,* and the commoners made use of public ones built by the rulers.[15] While people may have washed their hands or other body parts in cold water, bathing for hygiene, cleanliness, and ceremonial reasons took place in these *temazcales.* For drinking, clean water from the aqueduct was collected in canoes, or from the fountains, and sold by water merchants. In this water culture, extensive hydraulic infrastructures encouraged a wide range of uses and intimate daily contacts with water, an experience organized by elaborate ideas and concepts ranging from sophisticated knowledge of the qualities of different kinds of water and their effects on agriculture, to a deeply felt respect for maintaining the cleanliness of both their bodies and the lakewater around them.

BATHING IN THE MEDITERRANEAN WORLD

It is clear that the practices, meanings, and infrastructures of bathing in Mexico today are products of a long encounter between Europe and America, and water cultures in Tenochtitlán in 1490, on the eve of contact, were not the same as those in Andalusia. But while the conquest of the Americas was obviously an antagonistic meeting between people of two continents with no prior contact, to understand how the fusion of the two transpired it is essential to remember that on both sides of the encounter the peoples and cultures were already fusions of many earlier encounters. Furthermore, portraying Mexican water cultures as a colonial fusion of "Spanish" and "indigenous" directs attention away from the enormous changes that occurred between 1492 and the present day, in favor of the curation of hypostasized cultural survivals. Rather than cast Mexican water cultures in Mexico as the mixing, or *mestizaje*, of some fixed set of European bathing traits on the one hand and those of indigenous "deep Mexico"[16] on the other, I will start by showing how those traditions were already products of previous encounters. We have seen, for example, how the Mexica incorporated the engineering expertise of Nezahualcóyotl and other Texcocans in building Tenochtitlán, the city in the lake. In this section I argue that the colonial bathing encounter was shaped in important ways by the deep religious and cultural conflicts in Iberia during the fifteenth and sixteenth centuries.

The general contours of the culture of bathing in Spain and the rest of modern Europe and the Mediterranean were established by the Romans, who carried a standard set of practices and infrastructures throughout the Near East, North Africa, and Europe, which, long after the fall of that empire, continued to be reshaped and reproduced. The Roman bath included different rooms with hot, warm, and cold pools of water, as well as dressing rooms and steam rooms, and these different baths served different purposes in line with specific conceptions of human health and biology. Hot pools and steambaths were believed to open the pores of the skin and allow the transpiration of unwanted substances from the body; cold water closed the pores again. Under the advice of a doctor, bathing in the correct kinds and temperatures of water exercised a positive influence on the humors of the body, correcting for imbalances. Going to the bath, or bathing, could mean swimming or lounging in any temperature of water or soaking up the humid heat of the steam room, a wide array of different contacts with water that continue to define bathing today. The waters themselves were also varied, as Roman baths utilized both thermal mineral springs as well as freshwater sources heated artificially. The particular qualities of all these different waters were appreciated for their therapeutic effects, and mineral waters were valued as powerful curative agents.[17]

The Roman baths were social centers, and many of the activities of daily life were carried out within their walls. People gossiped, ate food, exercised and played games, had sex, relaxed, hatched plans, and carried out affairs of business and government. By bringing wealthy and powerful men together, the baths were settings for the consolidation of the patrician class. At its zenith of wealth and power the city of Rome counted more than four hundred bathhouses, and there were hundreds if not thousands more scattered throughout the empire. Each of the mineral water spas was known for the particular properties of its waters and their curative uses.[18] The sociality of bathing in the Roman world also involved religious or spiritual dimensions, and baths were dedicated to gods of both the Romans and those they subjugated.[19] Roman towns throughout the empire were built on existing indigenous settlements with springs that held religious and social significance. The Roman baths in Bath, England, for example, were named "Aquae Sulis" in dedication to Sulis-Minerva, a hybrid entity that fused the Roman god of wisdom with what was most likely a water deity of western England, on the far edge of empire.

Water culture in the Iberian Peninsula was not a pure cultural product, waiting to be carried to an encounter with "indigenous" bathing in the Americas, but rather a continually changing, multistranded "selective tradition."[20] The Arabs played an especially important role in this process, rebuilding and conserving many baths in Europe and the Mediterranean world built on hot springs. They were experts in hydraulic engineering, and during the High Middle Ages ("*baja edad media*": eleventh to fifteenth centuries) when they governed the Iberian Peninsula they constructed more sophisticated and remarkably more extensive urban and rural water infrastructure than had previously existed. Bathhouses were common throughout the Arab world, and this of course extended throughout Spain. Contrary to some popular ideas about the medieval period, bathing and bathhouses continued to exist in Europe and enjoyed a resurgence in the tenth to twelfth centuries.[21] The baths of Barcelona, for example, were founded by the Arabs long before they passed into the hands of the Christian nobility, while the baths in Gerona were founded anew in 1194. Córdoba, the capital of the western Caliphate, was said to have nine hundred baths for eight hundred thousand inhabitants.[22] In Spain, Arabs inherited and advanced the legacy of Greek and Roman literature, medicine, and cultural practices, conserving those works and codifying bathing culture in Islamic religious practice.[23] The Ottomans, who ruled much of southeastern Europe and the eastern Mediterranean between 1300 and 1900, also reproduced and reshaped bathing practices and ideas, built bathhouses, and generated further borderlands bathing encounters.

THE ECLIPSE OF BATHING IN SPAIN, 1500–1600

The selective tradition of bathing in Iberia was forged in a context of prolonged religious and political conflict between Muslims and Christians. Bathing in medieval Spain was supported by a body of medical literature that came from Greco-Roman and Arab traditions and reproduced the organization of the bath around the Roman model with the cold room, warm room, hot room, dressing room, and resting room. Within the medical model elaborated by the Roman physician Galen, to which most doctors adhered well into the eighteenth century, bathing was important for carrying off the remains of digestion, which formed one of six groups of things—called "non-natural" or "necessary" things—that were not intrinsic to human bodies. Bathing, exercise, and sex, all of which produced sweat and the emission of fluids, eliminated the remnants of these things from the body, and following Aristotle, balance and moderation was considered the correct way to deal with them.[24] Arab scholars such as Avicenna, who carried forth the Roman and Greek intellectual traditions in medieval Spain, delineated four

kinds of baths—freshwater, seawater, hot springs, and steambaths. Medieval doctors described the effects and uses of different waters, such as sulfurous or ferruginous (iron) springs, and the only proscription to bathing came from those doctors
who argued that very hot steambaths were dangerous because they disrupted
the humors. In Arab Spain there was a good deal of tolerance and coordination
among different groups to enable access by all to the bathhouses. In the baths of
Castille, women and men bathed on alternate days, with Jews bathing on Fridays
and Sundays.[25] Similar schemes to enable access to baths by different groups in
pluricultural societies were common in the Mediterranean world until the twentieth century, especially in bathhouses that utilized hot springs, which are by nature
singular and limited sources that cannot be multiplied or expanded.[26]

Then, in the sixteenth century, people in Spain stopped bathing. María José
Ruiz Somavilla has argued that understandings of cleanliness and bathing under-
went a fundamental adjustment at this time, due to two kinds of historical factors.[27]
First, Christians only recently finalized the long struggle with Arab rulers over
the Iberian Peninsula—the Reconquista—and distrust and hostility by Christians
toward Muslims and Jews generated over centuries was codified under Christian
hegemony as forced religious conversions, laws forbidding suspect activities, and
the policing of customs by the Inquisition. At the beginning of the sixteenth century bathing institutions passed from Arab to Christian control, as was the case for
the fifty bathhouses in Malaga that were given by the conquering Catholic kings
to the Church.[28] Soon, however, these bathhouses and the very practice of bathing came under scrutiny for they were linked to the customs of "infidels," who
according to the racial concepts at the time were compelled to bathe by "inherited
blood."[29] Abstinence from bathing, by this same logic, was evidence of Christian
ancestry and a badge of purity. Converts, or "new Christians," were banned from
working in the bathhouses in 1527, and by 1567 these attitudes toward bathing
hardened into a decree forbidding bathhouses and bathing in Granada.

Following the legal measures against bathing and bathers, people were brought
before the Tribunal of the Inquisition, tortured and punished, under accusations
of bathing or even for being too clean. The suspects were often women, and from
the declarations before the tribunal, it seems that what excited the imaginations
of Christian men was the combination of hot water and nudity.[30] Moorish men,
however, did not escape persecution for bathing. Bartolomé Sánchez, for example, confessed to bathing in 1597 and was imprisoned with loss of all property.
Miguel Cañete, a gardener, was tried and tortured in 1606 under the accusation
that he washed in the fields where he was working.[31] The rejection of bathing in the
sixteenth century, although enforced by capital punishment, was never total, and
bathhouses remained open in many parts of Spain until the prohibition of 1567.
Even with the prohibition, bathhouses in Andalusia remained open and bathing
in private seems to have continued or perhaps even increased in inverse proportion to the reduction of public bathing. Furthermore, accusations of heresy were

directed most often at those known or suspected to be Jews, Muslims, or recent converts to Christianity, and so bathing was not as risky a proposal for others. Most likely it was those unimpeachable Christians who were seen returning to the water in the early 1600s, bathing publicly, in groups, in cold rivers and streams. Although it was more acceptable and visible by that point, people probably never fully stopped swimming and bathing in rivers and springs during the second half of the sixteenth century, despite the prohibitory attitudes and decrees.

Sexuality and morality were an area of anxiety associated with bathing, and ideas of health were closely related to those about masculinity, custom, and nature. The health of an individual was maintained through balance and moderation in one's customs (*costumbres*), for customary behaviors were considered indissociable from an individual's "nature."[32] In this deeply conservative perspective, health was attained by avoiding excess (*deleite*) and disordered appetites, for balanced and moderate habits resulted in a healthful physiological character. Nature/custom was considered to be the best doctor, and the best remedies for ailments were to be found in nature and good customs. In this conceptual universe, bathing one's entire body by immersion in hot water or steam was easily construed as an extreme act and thus a problem. The virility of men, in particular, was seen to diminish from bathing, an idea that Ruíz Somavilla attributes in part to the idea that men had sex with men in bathhouses. The larger fear, rooted in a concatenation of moral norms concerning religion, sex, gender, and citizenship, was, as Fadrique Enríquez wrote at the time, that in the baths the soldiers of Christendom "would be made accustomed to luxury, delicate and vice-ridden, unhealthy … skinny, without virtue, cowardly and fearful."[33]

The second set of sociohistorical factors that were driving a slow reconceptualization of bathing and cleanliness among intellectuals had to do with the incipient formation of merchant capital and the spread of Renaissance thought. As European powers reached out to control far-flung empires, trade networks, and colonies during the sixteenth century, the merchants who made fortunes from this new global trade formed a social group that did not fit into the old regime of peasants, artisans, nobles, and church. While not nobles, the emergent bourgeoisie wielded the economic power to consume the array of commodities that were captured by global webs of trade in the early modern world. The increasingly important idea that men should maintain balance in their customs and not overindulge in food, drink, sleep, sex, and other pleasures can also be read as a warning to the new bourgeoisie, and to those who sought to emulate their customs.[34] At the same time, the medieval belief that social status was inherited through lineage—"purity of blood"—became more flexible and elite social status required more visible proof in the form of material culture and customs. Cleanliness was one area in which social, moral, religious, and class distinctions were established. While full-body bathing was unacceptable in sixteenth-century Spain, keeping one's hands and face clean took on an increasingly important role. The lightness of the visible parts of the body, maintained by washing, was seen as a sign of purity of blood, and placed the bather beyond reproach.

The abolition of bathhouses and many forms of bathing put doctors in a difficult bind. They continued to read and respect the foundational works of Pliny, Aristotle, Galen, and other classical and medieval scholars who recommended bathing for curing diseases and maintaining healthfulness, but these ideas were increasingly at odds with the political culture of the time. Doctors resolved this contradiction by arguing that the bathing activities of the Romans and Greeks had healing properties in antiquity but not in the present. Thus bathing actually did have benefits, but the damage caused by any abrupt change in custom was greater than the benefit that could be gained by adopting bathing practices anew. The long-accepted idea that bathing was good because it opened the pores of the skin and allowed for "exhalation" of unwanted substances, was turned around to argue for the threat of contagion from the environment entering through those same open pores. The malleability to the point of outright incoherence given to medical concepts so that they would correspond to the social field of forces in the sixteenth century prompted one scholar of the topic to characterize attitudes and ideas about bathing as "ideological."[35]

Changes in ideas about bathing were accompanied by changes in practices. Full-body immersion and steambaths were viewed with suspicion throughout Europe. Instead, people engaged in a more limited washing of the face and hands, as well as the practice of "dry bathing," which was the changing, and washing, of linens, rather than the body itself. "Dry bathing"—the washing of underwear, really—eliminated the body's "exhalations" that were captured by undergarments. Among the wealthy, undergarments became far more conspicuous during this time, protruding from sleeves and collars as a display of the hygienic customs—and social status—of the wearer.[36]

Eventually in the seventeenth century bathing came back into fashion and people—first commoners, then elites—went back into the rivers and the thermal springs. In Italy and France, the recovery of Roman and Greek texts stirred a rebirth of hot spring bathing among nobles, a practice that was copied elsewhere in Europe, and which also fomented bathing among the emergent bourgeoisie.[37] As intellectuals relearned the texts from antiquity about water and health, Europeans adopted the installations and aesthetics of Classical bathing, conserving the wide array of bathing infrastructure and practices—hot and cold pools, steambaths, showers—as well as the rich social, sexual, and spiritual dimensions inherited from the Roman baths and recast in the Middle Ages in Christian terms.

BATHING BY IMMERSION IN MESOAMERICA

Bathing would once again become acceptable practice in the seventeenth century, but in the 1500s bathing held deep and powerful connotations of sexual and religious danger and was closely monitored. And it was in this context that the Spaniards arrived in the Americas to confront far greater cultural difference and distrust than that which characterized the relation between Christians, Jews, and Moors in Iberia. Like writing, worshipping, and so many other activities carried out by the indigenous people of the Americas, bathing came under intense scrutiny in the New World.

Strange as it might seem, there is no evidence that anyone in Mesoamerica soaked in hot water before the Spanish conquest. It is not that there was no contact with water: on the contrary, bodily contact with water was a central part of daily life for Mexicans before and after contact, and the chronicles written during the early colonial period remark on the cleanliness and bathing habits of the indigenous people.[38] Díaz del Castillo in his tale of the conquest of Tenochtitlán told of the liberal daily bathing customs of the Aztec elite, and of the barbers who groomed Tlaxcalans. Documents from after the conquest suggest a number of ways that indigenous people washed with water. The indigenous authors of the sixteenth-century *Florentine Codex* depicted a person sitting by a pool and pouring water over his head and body with a gourd (see Figure 2). In hot lowlands areas such as Veracruz, Yucatan, and the Isthmus of Tehuantepec, people bathed and swam in rivers and other bodies of freshwater. In the Zapotec dictionaries compiled by Spanish priests, for example, there are words for bathing and for soap, which indicates that bathing was done for cleanliness. There was even a word that specifically denoted "waters to bathe in."[39] Gerónimo de Mendieta, a Franciscan missionary in highland Mexico in the mid-sixteenth century, wrote approvingly of the custom of mothers to bathe their children in the cold water of "streams, rivers and springs, first thing in the morning," which he maintained made them stronger, as Aristotle had said it would.[40] According to Diego de Landa, a priest traveling in Yucatan, "the Indian women bathed often with cold water like the men, and with little modesty, for they undressed and were naked at the place they went to fetch water."[41] Mayan women apparently "bathed a lot, simply covering themselves from the view of the men with their hands."[42] I have found no mention anywhere, however, of indigenous people soaking in hot water.

Despite everyday washing and swimming in cold water, waters were seen to hold dangers, and rivers and creeks held powers that many indigenous people feared.[43] This was most likely also true for hot thermal springs, for there is no mention by any of the early chroniclers of indigenous people bathing in these waters. It is possible that the practice did exist, but did not make it into the historical record. But there is also no documentary or archeological evidence that people bathed by immersion in waters they heated themselves; no pre-Hispanic bathtubs, for example. If it was a common practice we should expect there to be documentation: nude bathing by immersion in hot water was a particularly troubling activity for priests and government officials engaged in the struggle over the Spanish bathhouses in the sixteenth century, and if there were such activity in the Americas, it would certainly have captured their worried attention.

FIGURE 2. An Aztec man taking a bath. Drawing from the *Codex Florentino*, compiled by Bernardo de Sahagún, c. 1540. Granger Collection, with permission of Age Fotostock.

What makes the absence of any mention of bathing by immersion in hot water even stranger is that hot springs abound in Mexico. Juan de Cárdenas, a doctor born in Seville who lived in Guadalajara at the end of the sixteenth century, remarked upon the "great number of hot springs," and reasoned that their heat was derived from the sulfur that they contain.[44] Despite these remarks and all the mentions of the cleanliness and bathing habits of indigenous people in Mexico by the Spanish colonizers, there is no record that indigenous people in pre-Colombian

Mesoamerica went into those hot mineral springs, or that they were a part of indigenous medicine. This absence is especially striking, because there were hot springs in the parts of the Mesoamerican highlands that were most densely popu-lated, and these were put to use for bathing by Spaniards after the conquest. The principal hot springs in the Valley of Mexico were located on a volcanic island that jutted out of the lake east of Tenochtitlán, now a hill known as Peñón de los Baños. In conquest-period documents and indigenous *codices,* the island with the hot springs is described as an off-limits hunting reserve owned by Aztec nobility, like the forest of Chapultepec, without any reference to bathing or other use of the springs.[45] This is also the case with other well-known hot springs. *Codices*— pictorial documents produced by indigenous scribes during the contact and early conquest periods—mention the hot springs at Ixtapan de la Sal for the production of salt, not for bathing, and Oaxtepec, Morelos, was known for its royal Aztec botanical garden rather than its hot springs.[46] The mineral waters of Tehuacán, Puebla, which became important for their therapeutic properties in the colonial period, were used in pre-Hispanic times for irrigation and salt production, as were those of the warm mineral waters in Hierve el Agua, Oaxaca.[47] Inca rulers soaked in Andean hot springs, but there is no evidence of a similar use of hot springs in Mesoamerica.[48]

Even after Spanish contact there is little record of bathing by immersion in hot water or hot springs. Records point to bathing by Spaniards who, despite the prohibitions on using bathhouses in Granada, built a bathhouse at Peñón de los Baños and, by 1600, were building similar installations elsewhere.[49] The sole men-tion of indigenous practices of bathing by immersion in hot water comes from San Bartolomé Agua Caliente, a town in today's Guanajuato that was founded in 1541 by Fernando de Tapia, a Christianized indigenous leader who allied with the Spaniards in the conquest and colonization of the Bajío. His daughter, Beatriz de Tapia, is credited with providing, in her last will and testament in 1602, the land, springs, and money to build a hospital in San Bartolomé to serve "indios naturales y pobres," a project that was not completed until the late eighteenth century.[50] It is unclear whether the indigenous people in that region bathed in those hot springs before the Spaniards and their indigenous allies colonized the area, but if they did it is certainly strange that there is no mention of this practice by Spanish priests.

THE *TEMAZCAL*

Instead of soaking in hot water, indigenous people in Mesoamerica took steam-baths or saunas. In the same paragraph quoted above where he talks about cold-water washing, De Landa goes on to describe the less-common practice of bathing with "hot water and fire," which he says was done "for health reasons rather than cleanliness."[51] It is clear that this was not bathing by immersion, as De Landa lists two kinds of bathing: washing with cold water, and going to the steambath, or *temazcal.* Washing was more clearly aimed at cleaning the skin, while the *temazcal* was a therapeutic, medicinal, and spiritual activity with strong social dimensions. The unimportance of bathing by immersion in hot water, naturally or artificially heated, is directly related to the importance of the steambath in Mesoamerica.

The *temazcal* was an important part of life for Native Americans from the Pacific Northwest to Central America, save for the foragers of the arid lands of northern Mexico and the western United States.[52] In Mesoamerica—from about Nicaragua to the Tropic of Cancer—people bathed in smallish structures of masonry or adobe (often referred to by anthropologists as "sweatlodges") into which heated stones were placed. Water was then thrown upon the hot rocks to make steam. Sometimes the sweatlodge shared a wall made of volcanic rock with an exterior fire chamber so that the heat of the fire would pass through that rock to the bathing chamber. Water was tossed on that rock to create steam for bathing. Bathers would symbolically enter the underworld when they passed through the door of the *temazcal*, which in preconquest time in Mesoamerica usually displayed a statue of Tezcatlipoca, the god of healing and the underworld.[53] Other images of gods were also displayed, including that of Tocitzin, or Teteo Innan, sometimes called "grandmother of the *temazcal.*"[54] *Temazcal*es have been found in elite and everyday residences in the Mayan region built at least seven hundred years before the arrival of the Spaniards, indicating a deep history of bathing practices and beliefs.[55] *Temazcal* steambaths remain an important part of life in indigenous areas of Mesoamerica.

The *temazcal* was a ubiquitous and multifaceted institution in Mesoamerica that played roles in cleanliness, therapy, socialization, sexuality, religion, and agriculture.[56] The Spanish, however, understood Native American culture in terms of their own ideas of morality and decency, and they sought to banish sexuality, religion and magic from the *temazcal* in order to reshape it as a social practice dedicated to health and cleanliness. There are many laudatory mentions by the *conquistadores* of how well groomed the indigenous people were owing to their frequent washing, but the steambath was barely tolerated and particular religious and sexual practices associated with it were singled out as unacceptably offensive and subject to investigation and eradication. The assault on bathing and bathhouses during the 1500s in Spain was an assault on the religious, ethnic dimensions that did not conform to the ascendant Christian view of society and culture. When the popularity of bathing returned in the 1600s, it was no longer associated with religion and socialization among particular subaltern ethnic groups such as Moors and Jews, but rather with practices of health and cleanliness practiced by the nobility and emergent bourgeoisie, as well as those who emulated them. This turn to cleanliness and health was also imposed on the American *temazcal,* with one result being the loss of historical knowledge about other facets of bathing, and another the predominance of therapeutic uses.[57]

Sex in the bathhouse was the biggest concern, and what we know about the sexual aspects of using the *temazcal* comes from official prohibitions, condemnations,

FIGURE 3. A *temazcal*. Drawing from the *Codice Magliabechanio*. Source: Wikipedia Commons. https://commons.wikimedia.org/wiki/File:Codex_Magliabechiano_(folio_77r).jpg#/media/File:Codex_Magliabechiano_(folio_77r).jpg.

and persecutions. One text tells of "many naked indian men and women committing within [the bath] a great ugliness and sin."[58] In 1569 a priest penned a series of questions about sexuality in the *temazcal*, to be asked to Native Americans at confession: "Did you sin with any women [in the bath]? Was one of them your family member or someone you know … your sister or your sister-in-law? Did you by chance kiss a woman, holding her breasts, touching her, wanting her and coveting her?"[59] Another priest noted that the *temazcal* was "illicitly used by men with women, and men with men," surely a problem for those who, operating under heteronormative assumptions, tried to eradicate sexual encounters in the baths by separating men and women.[60] Despite the efforts of the church and government to banish such practices and limit the function of steambaths to health and cleanliness, *temazcales* indeed were, and would remain, spaces of unseen and unsanctioned sociality among different ages and sexes of indigenous people. An ethnographer studying the use of the *temazcal* in Chiapas today notes their continued association with sexuality.[61]

The *temazcal* proved to be an exceptionally strong institution, and as the colonial encounter progressed, the use of the *temazcal* extended into other racial-ethnic groups, including Spaniards who built private *temazcales* in their houses.[62]

Increasingly the concerns about bathing were framed as a problem of public order and health, as well as a problem of sin.[63] In 1646 the Crown created an office called the Royal Protomedicato, an official group of doctors and medical experts who were in charge of inspecting pharmacies and apothecaries, reviewing medical publications, examining and licensing doctors, and prosecuting customs and practices that contradicted scientific and Christian principles.[64] The Protomedicato had neither the responsibility nor the ability to oversee customs among indigenous people, and so mostly focused its attention on the Spanish and *casta* groups. However, the *temazcal*, which was strongly associated with indigenous culture although used widely in New Spain, was an important concern of the viceregal government and the Protomedicato in particular. And so, when the Royal Crime Office decried the *temazcal* for inciting men to engage in sodomy, the viceroy was forced to act.[65] The viceroy Conde de Monclova (1686–88) decided to keep the *temazcal*es open, but in the subsequent administration of the Conde de Galve (1688–1696) they were closed while two doctors, Ambrosio de la Lima and Joséph de Oliver, conducted a study to determine the social and medical dangers and benefits of this form of bathing.[66]

In the report published by the two doctors in 1692, their scientific opinion about the benefits of bathing was strongly informed by the idea that Spaniards and Indians were different races of humans with different physical constitutions. De la Lima and de Oliver concluded that the *temazcal*es were useful for the well-being of indigenous people, in particular, but also for Spaniards and *castas,* "whatever their color." That said, the doctors suggested that "for Spaniards, water baths would be more useful than *temazcal*es because white people have a more severe temperament" that would be "offended by steambaths."[67] This advice was informed by humoralism, a theory inherited from the Greeks and Romans which held that bodies—and here races of bodies—were characterized by different balances of blood (air), phlegm (water), yellow bile (fire), and black bile (earth), which produced the particular emotional and physical constitutions of individuals and races.

By reiterating the acceptability of bathing for health and cleanliness, and condemning bathing for social, sexual, and religious purposes, the 1692 study and others published later in the eighteenth century helped reshape quotidian bathing practices and water cultures more generally. Despite the 1692 vindication, and the spread of the steambath throughout society, it continued to hover between acceptance and prohibition, and was an ongoing object of concern for the colonial government. Around 1725, for example, *temazcal*es were prohibited in the indigenous pueblo of San Juan Teotihuacan, causing loud protest, and in 1741 the census ordered by the first viceroy Conde de Revillagigedo counted twenty-four *temazcal* bathhouses, double the permitted number.[68] The complicated mix of benefits from curation and cleanliness and dangers of sexuality and sensual exaltation, as well as the fact that the *temazcal* was by the eighteenth century an accepted activity that extended throughout all levels and groups of colonial society, mobilized constant patrolling of the practice.

CONCLUSION: COLONIAL WATER CULTURES

As they rebuilt and expanded Tenochtitlán, transforming it into Mexico City, the capital of New Spain, the Spanish elite slowly replaced the lacustrine system with one modeled on that which they knew back home. It was incremental change in many interrelated aspects of life: the environment, the culture, the economy. Floods devastated the growing city in the second half of the sixteenth century, prompting officials to embark on an enormous, centuries-long project to drain the Valley of Mexico. In carrying out this project they ignored and denied the uses and meanings given to the liquid by indigenous peasants who depended on complex wetland ecologies for their livelihoods, in favor of the notion that water was an input in production and a threat to a city that should not be wet.[69] Colonial public works extended those erected by the Aztecs to protect Tenochtitlán from flooding and provide freshwater, but they started from different cultural assumptions about the environment and the relation of humans to it and to one another. In confronting the peculiar environment of the Valley of Mexico the Spaniards were guided by a view of nature and humans inherited from scholars who lived in dry places—Hippocrates and Galen, Avicenna, and Pliny—and this view did not lead them to a harmonious and successful adaptation. In the process the Mesoamerican water culture that was relatively well adapted to the environment was dissolved, reworked, and transformed. As Alain Musset puts it, "the battle to control water was as cultural as it was technical."[70]

The very method of Spanish rule assured the continuity of indigenous water cultures, however. Like the Aztec rulers before them, the Spanish focused on controlling land and labor and extracting tribute. The indigenous peasant economy was the basis for the reproduction of the labor that enriched the Spaniards, and was left alone in many ways. The Spaniards did not try to eradicate hunting, fishing, and collecting resources, and alongside these basic economic activities, beliefs and ideas concerning water also persisted through the colonial period.[71] Both Spanish rulers and the indigenous ones before them made herculean efforts to keep water from flooding the island-city and to provide clean freshwater for its expanding population. The Aztecs built levees and raised the city up by filling in the lakebed; the Spaniards lowered the lakewaters by draining the Valley. The indigenous rulers built a kilometers-long aqueduct across the lake to bring clean freshwater from the springs at Chapultepec to the island-city of Tenochtitlán. The Spaniards adopted the same solution, rebuilding the aqueduct to channel new sources of fresh water into the city even as they drained water away from it.[72] This massive project to drain the Valley and bring in water was carried out by indigenous workers and the Spaniards learned from them.

Quotidian understandings of and engagements with water shifted as well. Indigenous people took steambaths rather than bathe by immersion, and hot mineral springs were apparently not utilized. The *temazcal* remained important in Mesoamerica, spreading as a practice through all social castes and classes, facili-tated, perhaps, by a long-standing familiarity with the Moorish steambath among the Spanish that was not entirely negative. But the holistic practice involving reli-gion, sexuality, and ideas about human and agricultural fecundity was narrowed through repression to concentrate on health and cleanliness. On the other hand, in the first century of conquest Europeans started the practice of bathing by immer-sion in hot water and, in particular, in the hot springs of highland Mexico. This form of bathing, and all its associated European ideas about health and the cura-tive properties of waters, also spread through the different castes and classes of colonial society. By the eighteenth century, both immersion baths and *temazcales* were common in the bathhouses of Mexico. These colonial baths and bathing practices were, in turn, subject to new forms of scrutiny and regulation in the age of the Enlightenment.

Policing Waters and Baths in
Eighteenth-Century Mexico City

Judge Baltasar Ladrón de Guevara walked through the streets slowly but nervously, attentive to the air of tension. After several drought years water and food was scarce in the capital, and he worried that in 1785 Mexico might see the kind of social unrest that recently beset Spain. He paused frequently to ask questions of people in the street, and so that the officials who accompanied him could record his comments and note the precise geographical coordinates of the issues he encountered. Returning many times to wander the same neighborhoods, he grappled conceptually with the diversity of peoples and *castas* that formed the city's immense underclass. It was an age of revolutions, and to deal with that instability Ladrón de Guevara was taking a new approach to government. His long walks and careful social study of Mexico City were aimed at understanding and improving the underlying organization of society and its relation to the environment.

Source: De Gortari Rabiela 2012: 122–23.

Enlightenment reformers like Diego Ladrón de Guevara were struck by the amorphous, variegated character of Mexico's underclass—the *plebe*—and its disorderliness caused them to worry. Mexico City official Hipólito Villaroel, for example, described this underclass as a "monster of many species." By 1785 the heterogeneity of colonial society overflowed existing legal, political, and socioeconomic institutions and concepts, prompting Ladrón de Guevara and Villaroel to engage in a systematic effort to comprehend it and order it anew. After a long period of careful fieldwork, Ladrón de Guevara designed a new administrative geography for Mexico City that he hoped would address social tensions. This was a new understanding of society and approach to governance that spread throughout Europe and the Americas in the eighteenth century and was known as "Police."

In this chapter I analyze ideas about Police in eighteenth-century Mexico City, and the implementation of those ideas to govern waters, baths, and bathing. Early modern treatises on government used the concept of Police to describe the discursive and institutional regulation of territory and population.[1] In a sense of the word that we would recognize today, Police was an institutional apparatus that operated in a prohibitive fashion to ensure security. Just as important, however, was another meaning: the wide, positive effort at civilized urban ordering that focused on aspects of the well-being of the population such as public health, the provisioning of food and water, and the maintenance of infrastructure. Surely the actions and ideas of Police were concentrated in the educated, usually European-born elite that occupied the heights of government in New Spain, but over time they shaped hydraulic infrastructure and were unevenly internalized and enacted by everyday people.

Ladrón de Guevara and Villaroel recognized that their ideas and practices of Police formed part of a material social field that included the environment as well as the environmental ideas and practices of the varied social groups they were tasked with regulating. As the eighteenth century advanced, new social groups emerged and consolidated in Mexico around socioeconomic activity, creole/national identity, liberalism, and science. I suggest we treat these intergroup relations as dynamics of class, and that we can see them in everyday conflicts over waters and bathing.[2] The history of these relations between "rulers and ruled" reveals that the policing of waters was always partial, selective, contested, and incomplete.

POLICING THE ENVIRONMENT

There was never enough potable water for the growing urban population in Mexico City. Beginning in 1548, the Spaniards rebuilt the aqueduct from the Chapultepec springs to the city center a number of times, and extended it to bring water from the springs at Santa Fe, high in the western mountains. This urban hydraulic system served the city as the principal source of clean freshwater for the next 350 years, and while the government carried out an enormous effort to drain the Valley of Mexico, the freshwater supply remained relatively constant. *Mercedes* (concessions) of water for houses and buildings were costly and hard to come by, especially as the city grew in the eighteenth century, and most people received water from the system of public fountains that were also served by the water distribution system. Water sellers, or *aguadores,* used these fountains to fill wagon-borne barrels and large jugs slung across their backs, and sold the liquid door to door.

The relative difficulty and costliness of delivering water made bathing by immersion at home impossible or impractical for most, and so people went to bathhouses, or simply bathed in the public fountains or the streams and canals of the lacustrine city. The latter options were of great concern among reforming elites whose sense of order was affronted by public nudity as well as the use of the same source for both drinking and bathing. Bathhouses were for bathing and washing clothes; fountains provided water for cooking and drinking. In this logic, the city's *temazcales* were acceptable to the city's rulers because they used relatively little water, and kept the act of bathing away from the potable water infrastructure.

The scarcity of water in Mexico City was exacerbated by natural events. The years 1780, 1782, 1784, and 1785 featured drought, which resulted in shortages in food, famine, and unrest across central Mexico in 1785 and 1786.[3] In Mexico City, a "severe lack of water in the public tanks," especially the north and east of the city center, lead to "exasperation and clamor among the poor."[4] In response, officials implemented scientific water policing measures that emulated those taken by the Crown in the 1760s in response to similar drought-induced bread riots in Spanish cities. The first step was a careful study of the environmental and social situation by the Maestro de Obras (Chief Engineer), who reported to the city council the leaks all along the aqueduct, the "deterioration of the pipes [most were clay] which are old and feeble with frequent breaks, the continuous leaking caused by the porosity of the lead [seals and joints], the *Mercedes* of water that continue to be delivered even after their titular owners die, and the excessive waste of the fountains that spill water."[5] The small reservoir (Alberca Chica) in Chapultepec was locked and the key was lost, and the large reservoir (Alberca Grande) was almost completely empty, its springwater flowing through an open gate down to the lands of the nearby *hacienda* of the Condesa de Miravalle.

Environmental pressures and enlightenment ordering coincided to propel the policing of new arenas of governmental concern, such as public health. Based on the report of the *maestro*, in 1788 the viceroy ordered all the *mercedes* of water to be registered and confirmed, and the springs and reservoirs in Chapultepec to be cleaned and repaired. Especially important was the main canal that delivered water from the springs in the mountains at Santa Fe to the aqueduct that began in Chapultepec. The *maestro* pointed out that the city's potable water often ran in an open ditch, and that people along the canal used the water for drinking, irrigating their fields, and bathing.[6] To prevent unacceptable use and contamination of the city's potable water by the bodies of its residents, the viceroy asked the *maestro* to cover the aqueduct and fence it off. Two years later the viceroy complained that despite these measures no positive increment in water levels was seen in the city, and ordered a report on the effects of the infrastructural changes. He furthermore ordered the *maestro* to build ditches to capture the runoff water from the Chapultepec springs and channel it back to the aqueduct that led to the city center.

The policing of the city's infrastructure included attempts to control bathing practices among peasants and urban poor, and worry extended to the ways people interacted with the liquid and with each other. The fountains in Mexico City provided water to most of the population, and were seen as an especially important site for the promotion of good health. In 1786 the viceroy Manuel Antonio Florez denounced the custom of the poor of using the water that spilled from the fountains and leaked from the pipes for bathing and washing, activities "not at all appropriate for the public streets" that caused "discomfort for pedestrians, complaints and risks."[7] Bathing and washing clothes in the canals and fountains of the public water system were prohibited and punished, which circumscribed poor people's access to and interactions with water. In 1790 the Conde de Revillagigedo tried to stop people from "bathing in the public canals" in Mexico City by ordering the construction of a public bathhouse.[8]

These enlightened government reforms were often contested. Fountains, like the springs and rivers that fed them, ran all the time, and much of the water went unused. They consisted of a spout that poured the water from some height into a pool or reservoir below, and their design was meant to allow users to take water from the cascading stream, rather than from the pool, which was considered unclean. In practice, however, the ranks of *aguadores* (water carriers) who serviced most of the population in the city did not wait to fill their amphora from the cascade, instead using the reservoir, which caused great concern among city rulers about the negative effects of this water on the health of the citizens.[9] To remedy this disorder, in 1790 the Conde de Revillagigedo replaced the central fountain in the Zocalo and its large pool of water with four smaller fountains, none with a reservoir. This measure, like many the Conde enacted in his period of reform, was rejected by the frustrated clotheswashers, bathers, and *aguadores,* who vandalized the new fountains and, in 1794, succeeded in having them dismantled.[10] The cleanliness and healthfulness of the water was not as important to them as plentiful supply and ease of access, but the destruction of the fountains ended up reducing their access to water. A few years later the neighbors in the barrio of San Sebastián complained to the city government that there was no fountain nearby, and so their family members walked a great distance to get water, which resulted in their sons getting into mischief and their daughters being "deflowered."[11] This and other complaints identified scarcity as the problem, rather than the unhealthiness of the water itself, which remained more of a concern of government officials.

Water policing featured strong technological dimensions. A 1792 article in a Mexican scientific journal identified the cause of "the hydraulic problem" in Mexico City to be the lack of circulation resulting from the city's location on a flat lakebed. In such a "horizontal" city water did not flow well. There was little elevation difference between one fountain and another, and so the public fountains could not be connected in a series. The unused water from fountains that ran twenty-four hours a day could not be efficiently channeled to other fountains, and instead overflowed onto the street and joined the wastewater and rainwater that coursed into the canals and finally into the lake. Surprisingly, the author found there to be an abundance of water: the Chapultepec springs alone provided enough water for a city four times the population of Mexico (home to about 213,000 people at that time). When measured together with the Santa Fé aqueduct, it seems there was water enough for millions of inhabitants![12]

This situation of both great scarcity and great waste was noted in 1786 by Ladrón de Guevara, who suggested that bronze faucets be installed, and taps and plugs be used, on all household fountains so that "no more than the amount of water necessary for the use of the houses and their neighbors would flow forth, avoiding in this way the spillage and lost water that runs out of the gutters into the streets."[13] The author of the 1792 article echoed this impulse to save water and offered plans for a machine designed to regulate the amount of water that flowed into the fountains. It was a valve, linked to a float, that would increase the flow when the fountain's reservoir was less full, and decrease the flow when it was fuller, shutting it off completely before the reservoir spilled onto the street. The circulatory concept that oriented the building of continuously flowing urban hydraulic systems was matched by a repudiation of waste, inefficiency, and shortage, and the design of mechanisms to restrict and administer those flows. Again, these enlightened measures to restrict supply were viewed with suspicion by the people whom they were intended to benefit. Anticipating that such policing of the environment would be rejected by the masses of users, the proponent suggested encasing the machine in a box to shield it from vandals.

POLICING THE BATHHOUSES

Bathing was a major concern of Police in Enlightenment Mexico City. Enlightened rulers sought to eradicate bathing and washing in public, and to move these encounters with water into bathhouses. In Spain, after a century in which bathing—especially social bathing—was discouraged, outlawed, and largely eradicated, people took to the water again in the 1600s. During that same period in Mesoamerica the conquerors repressed the sexual, social, and religious aspects of *temazcal* steambathing in favor of bathing for health and medicinal ends, a negotiation which enabled the *temazcal* as an institution to survive and spread across racial, ethnic, and class boundaries. Bathing in hot water was introduced by the Spanish into Mesoamerica, and by the eighteenth century bathing by immersion was firmly established as an acceptable activity both in Europe and the Americas, good for promoting a person's well-being and health. Moreover, bathing in hot springs had surged back into popularity, and the fashionable practice of taking waters percolated down from the nobility to the emergent bourgeoisie and other more humble social groups. Like the *temazcal,* bathing by immersion in hot water was considered therapeutic, and the mineral waters themselves were thought to be medicinal. As we shall see in chapter 4, the modern science of medicine and chemistry grew up around the study of these heterogeneous waters, and there was a surge of interest in Mexican mineral and hot springs at the end of the eighteenth century, stimulated by this inquiry and by an effort on the part of Spain to generate knowledge about its colonial territories, populations, and resources in order to generate wealth and govern more efficiently. Thus the increased prominence of bathing was accompanied by heightened policing of the activity. Just as the anxiety of Spanish priests moved them to chronicle *temazcal* bathing practices in the sixteenth and seventeenth centuries, the enlightenment project of policing the boundaries of the acceptable generated documentation that today provides a window onto the everyday practices of bathing that made up a central part of the water culture of the time.

By the middle of the eighteenth century the bathhouse in Mexico City was an accepted and commonplace institution with its own infrastructural and social requirements. Bathhouses usually offered facilities for immersion, steambathing, and the washing of laundry. They were located in houses that were large enough to accommodate a number of tubs (*placeres*) grouped in a single room or separated into individual stalls. Some bathhouses offered only cold baths, but in those that offered hot water there was a boiler in another room. The *temazcal* was a fixture in the bathhouses, occupying an open space such as a patio or courtyard where smoke from the fire could dissipate. The baths required a concession of water, or *merced,* from the government of Mexico City, a privilege which only larger buildings usually enjoyed. Houses, or institutions that grouped many people under a single roof, such as schools, hospitals, and religious orders, had *mercedes* of water. The *merced* allowed the building owner to install an intake pipe from the city's water distribution system, which then could be used to fill a tank or reservoir. *Mercedes* were limited in number and new ones were seldom awarded. Most inhabitants of the city did not have pipes in their homes and instead carried water from public fountains or bought water from *aguadores.*

Bathhouses were owned and operated in a variety of ways. Some were run as businesses, with the income derived from charging people to use the *placeres* and *temazcales.* Other baths were run as charities, especially by religious orders that tended to provide bathhouses for indigents, sick people, or rehabilitated prostitutes. It was especially common for bathhouses to cater only to women because it was prohibited for both sexes to use the bathhouse, and also because men could more easily bathe in the streams and canals that were a common feature of Mexico City and its outskirts well into the nineteenth century. The owner of the house where the baths were installed could be a private individual or corporate owner, and religious groups such as convents or orders of priests were especially prominent. Often the locale and *merced* of water was rented to the person or group that ran the bathhouse, with the same variety of lay and religious actors operating the establishments.

Every bathhouse was maintained by a set of required jobs that included tending the fire and the boiler for the hotwater baths, tending the fire for the *temazcal*, filling and emptying the *placeres*, carrying buckets of water to the *temazcal*, cleaning the installations, and handling money. There was often a bathhouse manager who oversaw the activities for the owner, and there was always at least one *temazcalero* (sweatlodge worker) to carry out all the menial labor. The sex of the workers was especially important, for it was inappropriate and outlawed for men to be in contact with women bathing in the *placeres*, and especially in the *temazcal*. Archival records give the impression that *temazcaleros* in the bathhouses of Mexico City were often indigenous people recently arrived from small towns in the Valley of Mexico or nearby who lived in the bathhouse and earned room and board along with their wages. In the case that a bathhouse was run by a religious order or convent, the work was performed by its members.

In Mexico City, bathing of all kinds became more frequent over the course of the eighteenth century, especially among the poor. According to the city government records consulted by the viceroy Conde de Revillagigedo in 1793, there were twelve bathhouses licensed in 1691 by royal decree, and in 1741 this number was increased to twenty-four in order to serve the growing population.[14] In 1741 Don Leandro Manuel de Gogochea received a license from the viceroy Conde de Juanclara to establish a new bathhouse with a *temazcal* for women in his house on the Calle de la Servatana, and then again in 1744 he was given another license to use his water concession to open a bathhouse for women in his house on the Calle de la Miserecordia.[15] In 1743 Leandro Manuel de Coxenechea y Carreaga received permission from the viceroy to open the Casa de Baños del Comercio at #22 Calle del Coliseo Viejo, also only serving women. More licenses were awarded in 1750 for bathhouses that offered *placeres*—or bathtubs—along with *temazcales*.[16]

The expansion of the practice of bathing in the eighteenth century also spurred the creation of bathhouses that were not licensed. In 1778, Sebastián Fabian and Miguel Pedro, *caciques* of the indigenous barrio of San Hipolito (just northwest of the Alameda), complained bitterly to the city government that Miguel Oballa purchased a house in that neighborhood with the goal of establishing a bathhouse. The problem, they stated, was not the bathhouse in itself, but the fact that Oballa lacked a *merced* of water and was therefore robbing the barrio's water system to supply his business.[17] *Temazcales* operated on unoccupied lands by rivers and canals, offering steambaths to those who used those bodies of water to bathe. The increase in licensed and unlicensed bathhouses in Mexico City shows that bathing acquired a greater importance in the eighteenth century among the poor, who, according to the second Conde de Revillagigedo, were "the people that use them most."[18]

The upswing in bathing was due in part to the increasingly accepted idea that it was an activity that should be promoted due to the health benefits it offered to both individuals and to the population as a whole. While plagues and diseases always caused fear and concern among rulers and ruled alike, it was during the eighteenth century that the more encompassing category of "public health" came into being in Mexico as an object of analysis and government intervention.[19] Bathing was a key practice of public health that quickly spread from the educated elite through the popular masses. For example, when applying for a license to build a bath and *temazcal* the owner of the property called "La Quemada" made it clear to the police department that they were supporting bathing because it was "public, and noted for its medicinal qualities."[20]

The increasingly common assumption that bathing was good for the health of the population arose from various roots. The *temazcal* was always viewed as therapeutic, and was used since before the conquest to remedy specific health problems. One of these uses was to help women purify themselves and recoup forces after childbirth, a practice that remained strong throughout the colonial period, as *temazcal* bathing was adopted into the institutional medicine practiced by doctors and hospitals. In Triptio, Michoacán, the hospital run by the Augustine friars utilized a *temazcal* in the 1540s, as did the Hospital Real de los Naturales in Mexico City, which cared for indigenous people and also made use of the hot mineral springs in Peñón de los Baños.[21] The Hospital del Amor de Dios, founded around 1540, featured, by the eighteenth century, a *temazcal* and bathhouse in its building on the Callejón del Amor de Dios. This *temazcal* provided therapy for patients, but was also a business that provided income to its owners. When, in 1788, this *temazcal* came under scrutiny by the city government, the overseer defended the steambath as a normal and accepted feature of any bathhouse in the city: "they all have what they call a *Temazcal,* which is looked upon as important medicine in the Capital, following general custom."[22] This was the colonial *temazcal,* purged of many of its indigenous religious and social meanings when it was adopted by the Spanish-dominated colonial society, and reshaped more narrowly as therapeutic and medicinal (see chapter 1). Another root of the idea that bathing was healthful came from Europe, where bathing and water were long associated with health and therapy, an idea that grew stronger in the seventeenth and eighteenth centuries.

The health benefits of bathing were, however, a promise fraught with peril, for bathhouses and *temazcales* were also known to be a setting for sinful encounters between men and women. The assumption was that these were either illegal sexual liaisons between prostitutes and clients, or simply the customs of Indians and poor mestizos that nevertheless offended God and the ruling class. In the Spanish tradition, social bathing was considered particularly dangerous for the honor of women, a guiding principle of gender present also in Latin America.[23] Such transgressions did not occur only among the unruly *plebe,* however. In 1779 a lower cleric (*racionero*) working in the Cathedral of Morelia, Michoacán, was punished for bringing his lover to the baths at the Cuincho hot springs managed by the Franciscan order. A short time later the viceroy received a complaint from the mother of two young women who were taken for a weeklong tryst to the same bathhouse by two other clerics.²⁴ Sex in the baths, by all accounts a rather common occurrence, caused a moral clouding of the beneficial waters.

Swimming showed the same Janus-faced character as bathing: widely accepted by the late colonial period as a healthy encounter with water, but still deeply suspect for its social implications. In the lacustrine city swimming in rivers and canals was a common social activity among the *plebe,* not easily distinguished from bathing. Elites also enjoyed swimming, however. In 1814, Don Manuel Pevedilla asked the Junta de Policía of Mexico City to award him a license for the swimming pool at his country house, which was used by his friends and acquaintances to "have fun." The pool was four feet deep, measured about sixteen feet by thirty feet, and was surrounded by walls on all sides. He assured the government that men and women would not swim together in his pool and that he would maintain order, but argued that it was the right of any citizen to have fun in his home. The government agreed to provide this permit for the pool, because of the sound moral and political character of Sr. Pevedilla, but also on the grounds that bathing in cold flowing water afforded proven health benefits. Ramón Gutiérrez del Mazo, the political chief of Mexico City who granted the permission, declared before the *junta* that swimming was a beneficial activity supported by wise policies and education, but that it was too often ignored or disdained in the heavily populated urban center. In his view, swimming should, like bathing, be promoted by enlightened government. The permit was granted under the agreement that men and women would not share the water at the same time: not even boys and girls could "have fun" together.²⁵ The fears of disorder, and the moral strictures about immersion and sharing water that those fears engendered, were centered on plebian bathing and swimming, but extended even to the private spaces of elite houses. Residents of New Spain learned to negotiate in their practices of swimming and bathing the tense opposition between the health benefits and moral degradation caused by these socioaquatic encounters, participating in this way in the disciplining process that was Police.

THE BANDO OF 1793

Together with improved administration of the *mercedes* of water and renovation of infrastructure, the viceregal government responded to chronic water shortages with a project to establish order in the use of public fountains and bathhouses. The famous modernizing viceroy of New Spain, the Second Conde of Revillagigedo, ordered a study made of the bathhouses, *temazcales,* and laundries of Mexico City, to establish the bases for an edict, or *Bando,* that reformed and regulated those establishments and the contacts that took place within them between people, and between people and water. The ordinance was proclaimed on September 24, 1792, and published in 1793. The *Bando* established rules to promote "public comfort, decency and health," part of a wider focus on "all the objects of policing in this

Capital." Public bathhouses were commonplace and bathing was viewed by the viceroy to be a "necessary" and indispensable practice that deserved his attention over other matters of state. It was not the first time the government issued rules for the bathhouses, but Revillagigedo complained that the government failed to hold the private bath owners and administrators to "the few rules already pronounced that favor good order and Public service."[26]

The viceroy, an American-born *criollo* administrator whose father served as viceroy of New Spain some forty years earlier, felt a deep commitment to colonial society and to changing it for the better. Revillagigedo brought a spirit of rational secular reform that was emblematic of the approach of the enlightenment Bourbon government, and he was notorious for his efforts at fighting corruption and enforcing the law. Similar in his governing style to Ladrón de Guevara, he proceeded systematically and scientifically in designing the new rules for bathing. First, the city sent architects and police officers (*celadores de Policía*) to visit all the existing bathhouses and evaluate their physical state, as well as the bathing practices that occurred within them. Based on these visits, the viceroy made an evaluation of the cultural traditions of bathing in Mexico City, as well as associated physical and infrastructural problems of bathhouses, before producing a long list of criteria that bathers, bathhouse owners, and employees were required to fulfill.

The 1793 *Bando* obliged all formal bathing and washing establishments to be licensed by the government, and established seventeen rules that *baños, temazcales,* and *lavaderos* were required to comply with. The measures were aimed at promoting "comfort, decency and public health" by stopping the "abuses, excesses and disorders that until now have reigned in the bathhouses to the detriment of the Public." The *temazcales* were viewed with suspicion by the governing Spaniards and *criollos,* and while the religious dimensions of their use were largely gone by the beginning of the nineteenth century, some of their social and sexual aspects remained. At a very basic level this was just a question of economy: the *temazcal* required a good deal of fuel, effort, and time to heat, so men, women, and children often used the *temazcal* together once it was heated. Certain *temazcales* were also most likely sites of sexual encounters, as they had been in the precolonial period, and the Revillagigedo government tried to eliminate heterosexual sex in the bathhouses by "cutting it off at the root." The neighborhood administrators of Policía were charged with enforcing the 1793 decree. Decency, or "good order," was paramount, and the first rule of the *Bando* was that bathhouses could only serve women or men, but never both. Unlike those earlier moments in the colonial encounter when the Church sought to eradicate the entire practice of the *temazcal,* the *Bando* of 1793 promoted this form of bathing while protecting a social and moral order based on the honor of women and a notion of public health.

Other rules aimed to separate bathers and create privacy within the bathhouse itself, a project that spatialized morality along the axes of caste and class. The *Bando* decreed that individual bathing chambers should be divided into separate rooms by unbroken walls, with no way to see bathers from the windows, or from the doors, if they were to be left open. The doors should all have locks, with keys in the possession of the bathhouse operator in case of emergency, as well as a straw mat to cover the floor and to rest upon, a bench or chair, and shelf for a candle. The *Bando* suggested that some, if not all, of the bathing chambers should have extra luxuries: a bell pull to summon the *temazcalero* to add water to the bathtub; hot and cold faucets to allow each bather to deliver all the water he or she should desire, rather than having it poured bucket by bucket by the *temazcalero*; a space for the bather's servant that was separated from the bathtub, again to maintain privacy. These rooms were for those who could afford them, while the poor used *bateas*—washbasins—housed in a single undivided room, "as was the custom." The individuation, privacy, and class distinction of the bather was achieved by these spatial and infrastructural regulations in the bathhouse.

Along with the *placeres* and *bateas,* many bathhouses installed *temazcales* as well as *lavanderos* for washing clothes, and these too were objects of policing that reinforced social distinctions. The *temazcales* were more popular among the poor and indigenous, and by 1741 there were twenty-four of these steambaths licensed by the government to operate in the city. Fifty years later, Revillagigedo's *Bando* allowed an unlimited number of bathhouses for bathing by immersion, but insisted on maintaining the number of *temazcales* at twenty-four. In that interim, many more unlicensed *temazcales* were set up next to canals and rivers that still existed on the city's fringes, serving the poor and indigenous people who used them most. The *Bando* suggested that all *temazcales* be located on the outskirts of the city "so that the poor people would have them closer at hand." The bathhouses of the city center, on the other hand, could focus on bathing by immersion, a prac-tice that required the running water of the public water system. *Lavanderos* were a series of washbasins in an open space and were used by poor women. Having studied the water culture of Mexico City, the viceroy found it necessary to prohibit these washerwomen from undressing and washing the clothes they were wearing, a practice that was common at the public fountains.[27] Finally, the bathhouses were required to have toilets, with cesspools or connections to the city sewers in the street that would carry off the human waste along with all the used water from the *placeres, temazcales, bateas*, and *lavanderos*. Shit, an unremarkable feature of early modern urban space, was recast as reprehensible.[28]

Police records about the enforcement of the *Bando* provide insight into bath-ing practices, social mores, and class dynamics in Enlightenment Mexico City. For example, in early 1793 José Molina, a neighbor of the "Padre Garrido" bathhouse, on the Calle de San Miguel, heard a ruckus coming from that business.[29] The neigh-bor happened to be the local watchman (*celador*), and knew that it was up to local police officials such as himself to uphold order and propriety in the city. The viceroy just recently announced new rules for bathhouses that were meant to eradicate the "disorder and disarray" that reigned in those establishments and to ensure orderly bathing for the benefit of public health. José, the *celador,* was compelled to investigate what seemed to be the sort of entrenched social bathing habits among the city's indigenous and poor that the new rules were meant to eradicate. The principal rule of this new *Bando* was that bathhouses for women such as this one were off-limits to men, so naturally José peeked through the door. He saw a large group of women and four men having lunch and drinking *pulque,* an alcoholic beverage favored by Indians. As if this was not enough of an affront to the civility of the public and the will of the viceroy, one of the women was undressing while the men were present. José dutifully reported this "disorder" to his superiors at the city police.

Witnesses were called to give testimony in the government offices. Standing before the police tribunal, the *temazcalero,* a "tribute-giving Indian" from Chalco named Lorenzo Francisco Antonio, identified himself and stated that because his boss, the female bathhouse operator, was gone at the time, he gave permission for the group to have lunch inside, but told them not to bathe until all the men had left. He stepped down and the next witness—the *temazcalero*'s wife, a mestiza woman from Mexico City named María Gertrudis González—was called to give testimony. As she rose to give her deposition, Lorenzo spoke briefly and quietly to María in Nahautl, with the Spanish police official listening attentively. María then proceeded to explain to the government officials that the group was accompanying a woman who recently gave birth (a *parida*) so that she could take the *temazcal.* It was understood by everyone in the room that such a visit to the *temazcal* was a common ritual in which relatives and friends participated, and was accompanied by food and *pulque.* The wife of the *temazcalero* finished by declaring that it was the husband of the *parida* who brought the buckets of water into the *temazcal*—not at all an indecent encounter. Last to provide testimony was María Antonia López, a Spanish woman who rented the building and operated the bathhouse. She placed the blame for the incident on the Indian *temazcalero,* saying that she was called away from her responsibilities because of a sick child, and that she did not give her employee permission to allow men and women into the bathhouse together.

When all witnesses finished their statements, the police official overseeing the depositions made a dramatic announcement. He had overheard Lorenzo Francisco murmuring instructions in Nahautl to María Gertrudis about what to say to the tribunal, and he would lock the *temazcalero* up in jail for "seducing and guiding" her. Later, when Don Bernadeo Bonavía y Zapata ruled on the case, he found the prisoner Lorenzo Francisco guilty of the "grave excess" of allowing men and women together in the bathhouse, and of allowing "scandalous abuses, entirely prohibited." He sentenced the *temazcalero* to eight days of hard labor on public works projects, cautioning him that there would be no mercy if such a thing happened again. The Spanish bathhouse operator, on the other hand, was simply cautioned not to abandon her duties again. Lorenzo responded with the formulaic utterance of the subaltern: "I hear and I will comply, but I won't sign because I don't know how to write."[30]

This courtroom drama tells us much about relations of inequality and power that surrounded bathing in Mexico City in the late eighteenth century. To begin with, while the *Bando* regulated bathtubs, *temazcales,* sinks, and clothes washing tubs, this and most other cases of transgressions and prosecutions only dealt with *temazcales.* The *Bando* reasserted earlier regulations that prohibited mixed-sex bathing in the *temazcales,* but said little about bathing by immersion. One judge explained in 1750 that the earlier rules were "provided by the Duke of the Conquest because there were too many disorders in those *temazcales*," and the regulation of bathing in the 1790s continued to focus on steambaths.[31] This shows that the regulations were an effort to change the water culture of the poorer, more indigenous social classes. *Placeres* were used by relatively affluent Spanish, creole, or mestizo people who could afford to conform to the moral standards and values of Enlightenment officials, such as privacy. Bathing by immersion in tubs of hot water was a relatively recent import to Mexico, and did not have the same deep religious, social, and sexual dimensions associated with the *temazcal.* The 1793 *Bando* certainly prohibited mixed-sex use of the *placeres,* but it influenced bathing by immersion in a more positive, rather than punitive, way through the architectural requirements it established. Clients who could pay for expensive *placeres* were to be provided with privacy and comfort: a vision for how bathing should be rather than an attack on what it should not be.

Bathing shows us how the class struggle between rulers and ruled was organized. Both the group that carried out the postpartum bathing ritual and viceroy's capillary police organization were motivated by concepts of cleanliness, decency, and public health, but the concepts held quite different meanings for these different people. This new definition of bathing proposed by the viceroy in the *Bando* of 1793 was clearly not shared by many of the clients of the bathhouses. There was an abrupt social divide based on notions of class, race, ethnicity, honor, and decency, and on access to one or the other form of bath: *placer* or *temazcal.* The *placer* was a tradition with origins in Spain, not the Americas, and bathing by immersion was imagined in the late eighteenth century as the more refined, European form. The *Bando* aimed to refine the *placer* even further, individualizing and privatizing the bathing experience. The poor and indigenous, on the other hand, conserved practices of ritual *temazcal* bathing in groups: in the case presented above, for a woman who just went through childbirth. This bath was a social ritual celebrating the birth of a child and the survival of the mother, who was accompanied by family members and friends to support her in the care of the infant as well as in the bath itself.

Police was an exercise of rule on the terrain of culture. It was not ignorance of customs that drove the viceroy to outlaw them in the *Bando,* but rather knowledge about them gained from careful study. The delegate of the Crown to rule New Spain imposed a concept of the correct way to bathe that sought to change popular bathing customs, obliging the mass of city residents who utilized public *temazcales* to do so quietly, orderly, individually, and in a way that repressed sexual and social dimensions in favor of emergent concepts of cleanliness, decency, and public health. Washerwomen, cleaning their own clothes as well as the clothes of their wealthier employers, were also singled out as a particular threat to the new order. They were admonished for nudity while washing; they were castigated for washing in the public fountains; they were accused of open defecation in the bathhouse.[32]

Cultural attitudes toward bathing do not derive in any automatic way from the social position of an individual, and in some cases it was young indigenous women who lodged complaints with the police that a man was attending to their *temazcal*. Bruna Cisneros and her sister María, both indigenous women from Mexico City, testified against José Anselmo Escobar, *temazcalero* in the bathhouse of the Calle de las Moscas, for entering the *temazcal* to make steam by throwing water on the hot stones and to pour buckets of water over the bathers. María declared that her sister was deeply ashamed because the man saw her body, and it was this shame and sense of honor that motivated her to report the breach of the law to Molina, the local police officer. While they were at the bathhouse to take a shared *temazcal*, by no means did these poor indigenous women feel comfortable with unknown men seeing them naked or entering into the *temazcal* with them, which reflects the limits of the sociality of bathing among subaltern folks in Mexico City. Although the sense of honor and shame articulated by María was rooted in traditions that went back well before the Enlightenment, it was also at the heart of a modern sense of public decency and order expressed in the *Bando* and articulated by her testimony and those of other subaltern actors.[33]

The records about policing the bathhouses in Mexico City shed light on both the emergence of modern governance and on the effects of policing on the ways people related to waters and to each other. The infractions to the *Bando* of 1793 judged by the *regidores* were brought to their attention by citizens who were local agents of Police—the *celadores*. José Molina, who took the initiative of investigating the "Padre Garrido" bathhouse described above, was from the neighborhood. The proclamation of the viceroy was enforced by the local agents of government who introduced the careful gaze of Police into people's neighborhoods, businesses, homes, and baths. Another police officer, Onofre Ramírez, sent his wife of twenty-five years to take a *temazcal* in a women's bathhouse he suspected of having male *temazcaleros* that mistreated the female clients. By her own account, Ramirez' wife was lucky to have escaped unharmed by the *temazcalero,* and remained fearful of retribution for her testimony.[34] Considering the risks, the commitment of both husband and wife to uphold the law and police the culture of water is remarkable. We can discern in these stories how an awareness of the rules and a consciousness of the position of one's self in relation to the rules were insinuated into everyday relationships, even the most intimate ones.

CONCLUSIONS

Water scarcity and social heterogeneity threatened to burst the levees of colonial order in the late eighteenth century. The city was imagined by its rulers to work like a human body, with its interrelated organs and circulation of fluids, and so government was seen to require the creation of infrastructure and the ordering of space. Society, too, was expected to be orderly, and government in the late colonial context was directed at managing the frictions between heterogeneous groups defined by origin, status, and *casta*. In late colonial Mexico, *policía* was the word given to this regulatory activity and institution by the viceregal government, a meaning that shifted toward "security" as the nineteenth century progressed.[35] A series of dictums issued by the Crown in the 1780s created institutions, codified policing, and heightened its importance to colonial administrators, especially in Mexico City.[36] Police encompassed the ordering of political economy and the control of urban space through the building of circulatory infrastructure and the management of wastes. It was, moreover, a moral project to manage populations by reforming behaviors and cultural practices and eliminating vices.

Public health was a key arena in which governing officials sought to expand the purview of policing. In the eighteenth century the health of people was considered to be intimately connected with the environment, as it had been since Hippocrates penned *Airs, Waters and Places*.[37] From this perspective, odorous airs, or miasmas, were held to be vectors of disease, and cold and heat were also blamed for health problems.[38] The watery urban environment of Mexico City was thought to produce these smells and airs, so public health measures tried to improve circulation of these substances and keep waste out of them. Good health depended on clean, constantly circulating water, and the government did what it could to keep the water in the city's pipes and canals safe from the polluting contact with humans and animals, and to keep it apart from the compromised waters of the lake and the rivers. It was this conceptual link between environment and health that gave rise to the thriving fields of climatology and mineral springs medicine in the eighteenth and nineteenth centuries.[39]

Water management was framed in terms of the health of the entire undifferentiated public. However, the issue of provisioning specific different groups of people in the city with liquid was never far from the surface, and much of the archival record reflects contention over the resource. Water was crucial, of course, to the livelihoods of peasants who depended on the lakes and irrigation systems for food. But water was also fundamental to the survival of the swelling urban populations of the eighteenth century. When urban water sources dwindled, or became contaminated, unrest followed. There never seemed to be enough water in the public infrastructure, especially for the expanding, peripheral neighborhoods of poor recent immigrants to the city, and so the policing of water was aimed at ensuring supply and access to the liquid for heterogeneous groups of poor and marginal people. Water management was fraught with peril, but its positive promise could be realized by building infrastructure and shaping ideas and actions.

A major focus of policing in Enlightenment Mexico City was the practices and infrastructures of bathing. After the sixteenth century during which bathing was viewed with suspicion, it regained by the eighteenth century a privileged place among civilized customs due in part to the success of the Spanish rulers in reshaping the practice. The *temazcal* had been largely purged of its indigenous religious and sexual associations and was by that time used by all social groups for cleanliness and health. Bathing by immersion was also on the upswing, influenced by similar shifts in Europe and North America and the encouragement of Enlightenment governments. The sciences of medicine and chemistry were growing quickly in the service of public health, and they focused on water and the positive improvements that could be achieved through its management.[40] Despite these changes, bathing retained worrisome moral and civic dangers. It cleaned and healed, but was also still a setting for sexual and social encounters that the government, as always, strived to prevent. Bodies were separated; men and women kept apart. Bathing by immersion was increasingly individualized and bathing in the *temazcal*—always a group activity—was segregated by gender.

While the contours of this overall shift in water culture are clear, a close look at the archives shows us that would be a mistake to award too much coherence, or effectiveness, to the project of Police. Far from a steamroller of spiritual history that functionalized people to the demands of bourgeois society, Police in Mexico City was a series of declarations, actions, and decisions, not always interconnected, that percolated unevenly through the urban underclass. The institutions of Police were similarly incomplete, with partial coverage and selective application. What policing does show is how the daily frictions between rulers and ruled were organized along lines of race, class, sex, and ethnicity, and how universalizing concepts such as "nation," "public," and "citizen" were deployed by the government and reworked with differing content by people according to their position within specific fields of power. Enlightenment water governance was an incomplete and fragmented project that would nonetheless gain strength over the following two centuries.

4

Enlightenment Science of
Mineral Springs

Reforming elites in Enlightenment Mexico City did what they could to clean up the disorder they perceived around them. Policing was their response—the rational management of populations and resources to ensure that both prospered. Studies were carried out to provide information useful for management of the varied castes and classes in the urban center, and for channeling waters more efficiently through infrastructures to private buildings and public fountains. The objects of reform were complex systems that melded resources, infrastructures, and human bodies, ideas, and practices; washing and bathing in particular were subject to scrutiny as cleanliness came to the fore as a pillar of public health and social order. Partly because of this complexity, the policing of waters and baths was haphazardly and selectively enacted.

The promise of bathing was not only moral and physical purity, however. For millennia waters were considered regenerative, therapeutic, and medicinal, and it is hard to overestimate the importance waters held for ideas about health in the ages before antibiotics and surgery. Waters were thought to both cure illnesses and prevent them. There were many categories of waters, each defined by a characteristic: salty, iron, soda, hot, warm, etc. Certain kinds of waters balanced the body's humors in certain ways; others were prescribed for skin problems, venereal disease, kidney stones, even madness. Waters were applied in an empirical and experimental way, based in the traditions passed down by healers and from the texts of antiquity.

In this chapter I continue the discussion of enlightened bathing by turning to the ways that the diversity of Mexico's mineral spring waters were studied, valued, and used during the late eighteenth and early nineteenth centuries. This is a story of the growth of scientific knowledge about chemistry, medicine, and related topics such as botany and human physiology, and the importance of springs in the development and application of this knowledge. Historians have shown that northern European mineral and hot springs were especially important sites for growth of science and medicine in the modern period, and this is also true for New Spain, what is today Mexico. A look at the colonial realm of the Americas, however, reveals social dimensions of water cultures that are not commonly portrayed in the literature concerning European hot springs, specifically issues of race, class, access, and power.

There is also something to be learned from American mineral springs about the particular development of science in those longitudes. Colonial science during the Enlightenment grappled with reconciling universal humanism and a search for the exotic and incommensurable. In other words, hot springs and the humans who used them (like plants, animals, and the rest of the natural world) were seen to fit into global classificatory schemes, but inherited European notions that the Americas were fundamentally different lingered. Expeditions were mounted to identify this American exceptionalism and incorporate it into the expanding classification systems of modern science. Scientists traveled to the far corners of New Spain in search of hot springs and measured their temperature, smell, taste, color, density, and chemical composition. These springs were usually already somewhat developed by local users, and local bathing and drinking customs were also chronicled by scientists interested in the medical applications of the waters and the possibility of developing them into spas like those that were become increasingly fashionable across Europe.

A close look at the Enlightenment science of waters reveals a key ontological difference from today's scientific understanding of them. Waters *acted*; they were medicinal, with qualities described at the time as "virtues." Most people now think of water as an inert, uniform liquid that is controlled and used by people to grow food or flowers and wash dishes, cars, or bodies. Scientists today largely share those ideas, although they recognize that the water molecule is polar, and thus can dissolve many substances, and that as a liquid it can erode solids. In the eighteenth century, however, different waters were seen to have other kinds of effects, and to have them on human bodies. Long before the recent appreciation for "vibrant matter" and nonhuman "actants" in anthropology, Enlightenment science strove to understand the powers of water, and how it formed assemblages with bodies and cultures that came together in human health.[1]

THE VIRTUES OF WATERS

From the early days of the colony until the Independence struggles, the Crown periodically carried out surveys and censuses of its territories and populations, both in Iberia and the Americas, for use in writing descriptive geographies called

*Relaciones Geográficas.*² A review of these surveys provides an overview of how, as the centuries progressed, the topic of mineral waters grew in importance, and also how the cause for that interest changed. At first mineral springs were important as sources of sodium chloride—table salt. Throughout Mesoamerica mineral waters were used for the production of salt, a crucial complement to the largely vegetable-based, sodium-scarce indigenous diet. Ixtapan de la Sal, a hot spring town in today's Mexico State, was famed in the Aztec empire as the source of clean, white salt that circulated as a tribute good. *Tequesquite,* a naturally occurring combination of sodium chloride and sodium bicarbonate that forms as encrustations on the soil, was also crucial for the diet and economy, and was used in its natural form for cooking and also processed to create sodium chloride. Inhabitants of the Valley of Mexico continued to produce salt from encrustations on the shores of Lake Texcoco until the early twentieth century.³ Salt was used for industry as well, such as the curing of meat and, after the arrival of the Spaniards, the refining of silver. So central was salt to the silver industry that in 1580 the Crown created a royal monopoly over its production and commerce, and while salt mines eventually provided the greatest supply, in the early years of the colony salty lake and spring waters were a key source of that mineral.⁴

The waters of the realm were valued for their "virtues": for their characteristics and efficacy in the world. In 1554, for example, the Crown asked local governments in Mexico to identify "big lakes or notable springs whose waters have some particular virtue" or usefulness due to some essential characteristic of that water. These virtues and their agency were not human, and "virtue" does not refer to the moral principles or forms of reasoning that philosophers in the Western tradition have long debated. Rather, virtues were qualities of the waters identified by their material effects on other bodies, human and nonhuman, and grouped by those effects and their assumed underlying causes. As time passed, the virtues of waters were increasingly defined in terms of therapy and medicine, and the long medical tradition coming down from antiquity through the Arabs identified categories of waters by their effects on human bodies, such as aiding in rheumatism, healing skin disease, or dissolving kidney stones.

While bathing was largely frowned upon during the sixteenth century, its medical applications grew to be accepted by 1600. A survey questionnaire from 1604, for example, asked that local government officials provide information about "medicinal springs or baths."⁵ During the eighteenth century interest in medicinal waters flourished, and the refinement of the categories to describe the variety of those waters is reflected by the 1777 questionnaire, which sought information concerning "hot waters, salty waters, bituminous waters, and those waters useful for certain illnesses, as well as the temperature of the waters and their respective bitterness, bituminous flavor or saltiness."⁶ By 1812 this interest blossomed into an entire field of questions in the census questionnaire (article 11) concerning "mineral waters and baths." Respondents were instructed to provide information about the waters themselves, as well as the practices of the people who made use of the waters, and the infrastructure that existed to support the people who visited those waters.[7] Underlying this inquiry into the utility of waters was the idea that waters were efficacious, agential, "virtuous."

The changes over time in the questions posed by the government about mineral waters reflect developments in science, medicine, and the business of bathing. Historians suggest that in the eighteenth century intellectuals began to approach waters and their medicinal qualities in a new way. Departing from the Hippocratic and Galenic tradition that was empirical and experimental, scholars working in the emerging scientific paradigm developed a theoretical and systemic approach to understanding the diversity of waters and their particular effects on human organisms and diseases. The results of these investigations into waters and bodies were applied to a growing model of the relations among substances in the universe—chemistry—and scholars distilled, processed, and analyzed the contents of mineral springs to identify their components, locate them in relation to other substances, and discern their effects on the human body.[8]

Mineral springs chemistry had great implications for medicine. Doctors brought new information about the substances in plants and waters to bear on existing schemes for understanding disease and well-being such as temperature, climate, and humors. In 1788 Mexican doctor Juan Manuel Venegas dedicated a section of his *Compendio de la Medicina* to water treatments, in which he discussed the effects of different temperatures and kinds of waters and different techniques of bathing. The hot mineral waters of Mexico were grouped into categories depending on their principal substance—ferrous, sulfurous, calceous, acidic, and nitrous—each with its applications to particular conditions. He listed dozens of hot springs in New Spain that had already been analyzed for the purpose of promoting medicinal uses. In addition to these hot mineral waters, the doctor described the efficacy of bathing in "common water" for certain conditions, which reveals both the conceptual existence of homogeneous modern water, and that it was at that time simply one kind of water among many. Doctors also elaborated a range of physical applications of waters to bodies—techniques of bathing that corresponded to particular waters or ailments. Half-body baths from the stomach down, or *semicupios*, were recommended for colic, kidney stone pain, and inflammations in the belly. Foot-baths were prescribed for headaches, facial paralysis, sideaches, and hemorrhoids. *Temazcales* were useful for sterility in women, paralysis, and "coldness" in the body. These applications were believed to compensate for an excess of some condition in the body (heat, cold, humidity, dryness, viscosity) with an opposing quality (temperature, chemical, area of application) of the water or its application. But while medicine and chemistry developed a new scientific paradigm for understanding waters, their heterogeneity and classificatory order was rooted conceptually in their virtues.

EXPEDITIONS

Most of what we know about the development of science occurred in northern and western Europe, and this is especially true of the science of waters. Although less has been written about colonial science, intellectuals in the New World were keenly aware of the latest advances and actively participated in these discussions.[9] Expeditions mounted in Iberoamerica in the late eighteenth and early nineteenth centuries sought new plants, animals, mines and resources, new routes, and unseen human cultural diversity. Occasionally these expeditions were directed to studying hot springs, and such waters were always mentioned if encountered en route. Colonial space served as a special sort of laboratory, where scientists mounted expeditions to search for the exotic, unusual, and exceptional, and to measure, analyze, and classify these novelties using new methods. The search for the exotic was an older project that lingered on after the paradigm shift to experiment. In 1591, for example, Juan de Cárdenas revealed to readers the "marvelous secrets of the Indies," recounting tales of petrifying waters that caused leaves and other objects to develop a stony, mineral surface. As we shall see, Enlightenment scientists brought new techniques of measurement and description to bear on similarly exceptional natural and human phenomena in the hot springs of New Spain.

Expeditions were controlled by the church and Crown until the early nineteenth century, when imperial power waned and individuals from other countries were granted permission to carry out studies of the Americas. Around 1770 the archbishop of Mexico, Francisco Antonio Lorenzana, sent Fray Pablo de la Purísima Concepción Beaumont to study the hot springs of San Bartolomé, near the city of Querétaro. The springs had been used since at least the 1500s by locals to relax and feel better, and by the late eighteenth century were managed by the Hippolyte religious order. In 1757 a priest from Querétaro cured his arthritis by bathing in the waters, which spread the fame of the hot springs among the urban elite. With the increasingly popularity of hot springs among bathers, scientists, doctors, and government officials throughout the 1700s, the church sought to develop San Bartolomé hot springs into a hospital and bathhouse. Beaumont had a degree from a university in Paris, and was well acquainted with the "particular treatises" concerning mineral springs in Europe and their curative effects.[10] His 1772 study of San Bartolomé was the scientific justification for plans to build baths and a hospital for indigenous people and to establish professional medicine at the site.

Beaumont sought to replicate the European model of the spa in rural Mexico. But despite this universal application of medicine and water science, he was well attuned to the specificity of both the springs and their users, for the attractiveness of San Bartolomé resided in its unique virtues. His analysis of the water begins with simple evaluations of its smell and taste: a light sulfur odor and a sharp taste of iron. Reduced by boiling, the water had an effervescent quality, and the solid residues burned readily when put to the flame. Confirming that the distilled water was "crystalline" and pure, he continued to process the residual solids with a variety of techniques, deducing finally that the waters were rich in "sulphuric, alkaline, fixed salts."[11] According to Beaumont, this combination promised to "dissolve thick humors" (the alkali) and serve as a sedative and balm for skin disease, respiratory problems, and paralysis (the sulfur). He prescribed the springs for treating arthritis, rheumatism, and gout, and described a set of bathing practices that must be followed.[12] The patient should take two baths a day of fifteen to thirty minutes, one at 10 a.m. and one at 5 p.m., and after each should be wrapped in clothes to promote profuse sweating.[13] Drinking the warm water would also induce sweating, which was considered the key to achieving results. For the treatment of renal or pulmonary problems, patients should take *semicupios* (half-baths) from the waist down. Other maladies required showers—water poured from a considerable height from a bowl.[14]

Beaumont also carried out a social study of the springs and their uses. The waters were used for all purposes by the locals. Of course they bathed in them, but having no other source they also drank only mineral spring water and suffered no obvious deleterious results as far as Beaumont could tell. Once the water left the baths, it irrigated fields of grain and vegetables, also with no negative effects on plants other than fruit trees, which did not grow in the area. The earth around the springs was saturated with minerals, and the Indians use it as a soap they called *Xaboxay* to wash their clothes in the mineral waters.[15] The spring itself was fenced off, and the waters were conducted from the source to a large pool through ceramic pipes. Beaumont considered this a wise design, for it prevented the popular practice of cooking chickens and corn in them, or using them to scald and pluck butchered animals. It also conserved the heat, and with it, many of the health benefits of the water. The priest viewed with distrust the practice of local men and women to bathe "one in front of the other, with their unclothed flesh exposed to the four winds," but recognized that because nudity "was almost a custom in these lands" that "maybe there is no spiritual danger."[16] Nevertheless, "some women who are not Indians take the shameful liberty of bathing in public," a practice Beaumont was eager to put an end to. He suggested that two pools be maintained, so as to separate the sexes and maintain decency and order.

The double standard by which Beaumont evaluates the morality of indigenous and nonindigenous bathers reflects an unresolved tension in Beaumont's treatise between his analysis of the universal benefits of hot springs bathing for humans and his proto-racial theory of the distinctiveness of different groups of bathers. The Indians of the region, he said, "almost lived in the San Bartolomé hot springs, bathing there at all hours of the day and night."[17] This was natural, he reasoned, for their work in the fields gave them an excess of cold and humidity in their bodies, and the hot springs were the remedy at hand. They "live always naked," which made their skin less delicate and more resistant to the heat of the water. But beyond the cultural differences, Beaumont believed that Indian bodies had a particular physiological and chemical composition which responded especially well to these waters. Their bodies, he wrote, were "very oily, their sweat is thick, which is why they do not get gray hair until very old, and, as I have observed in the Real Hospital de Naturales in Mexico, their bones are full of Sulphur."[18] He maintained that the bones of Indians were "spongy, filled with lots of oily marrow, and sulphurous," which allowed them to bathe at length in the San Bartolomé hot springs and extract great benefits from it. Just as different ailments responded to a water in different ways, so too did different bodies. This understanding of bodily heterogeneity and variable "virtue" jostles alongside Beaumont's framing of the study as an effort to serve a universal "public good"—the health of the population.[19]

This tension between the heterogeneous and homogeneous was at the center of many efforts to understand the world from the emergent scientific perspective in the Enlightenment. Antoine Lavoisier, who identified a number of elements and contributed to the elaboration of the periodic table of elements, was particularly influential among intellectuals in Spain and Mexico. Even before the publication of his major work in 1789, the idea gained traction that water was a pure substance composed of two hydrogen and one oxygen, and that dissolved into it were other substances that provided all the waters of the world their particular properties. Earlier descriptions of diverse waters based on geography, temperature, astrology, and supernatural forces were replaced with a Linnean array of categories based on the principal impurities found in water: iron, sulfur, carbonate (soda), salt, etc. As Jamie Linton (2010) summarizes it, this was the period during which "waters" became "water," but the creation of the singular did not eradicate the existence of the plural, and hot springs waters continued to be understood in terms of their particular yet variable virtues.

The system of intendancies—regional jurisdictions—that was established in Mexico by the reforming Bourbon government in its *Ordenanza* of 1786 obliged regional governors to tour their regions and collect scientific information concerning natural history.[20] This information was expected to help the governors to carry out their duties in the area of Police, and to marshal the resources of the colonies for political and economic development. In March of 1789, the governor of the Intendencia of Valladolid (today Michoacán), Juan Antonio de Riaño, was surveying his territory together with a group of German engineers, in the hope of developing mining resources. While exploring the Jorullo Volcano in Michoacán, they stumbled across a ravine on the flanks of the mountain where there were "various hot springs used as a bath by sick people."[21] Riaño expressed concern that there were no scientific analyses of the waters, nor doctors to oversee their use by locals. "It turns out," he reported, "that most of the time people bathe in those waters who have diseases that are of such a nature that they do not receive any cure, but rather considerable harm."[22] He collected samples of the water with the idea that scientific study by colleagues in Mexico City would promote the development of a medical spa business in the Jorullo hot springs. These brief experimental forays of institutional science had little effect on local water cultures, and people continued to use hot springs in accordance with their inherited customs. Nevertheless, as we shall see in subsequent chapters, the struggle over ideas and access evidenced in Jorullo defines much of the history of Mexican hot springs from the Enlightenment onward.

With both scientific and economic goals in mind, the Crown financed a series of expeditions in the Americas that collected information about natural resources such as plants, waters, and minerals in order to develop their medicinal uses. Hot mineral springs received special attention because all three of these aspects could be studied at the same: the waters, their mineral content, and the plants that grew around them. At the same time, a series of laical scientific institutions were created in Mexico: the Royal Academy of San Carlos (1781), the Royal Botanical Garden (1788), and the Royal Mining School (1792). The growth and institutionalization of a scientific community of pharmacists and doctors, and the lively discussion of the ideas of Linnaeus and Lavoisier that ensued, helped motivate expeditions to collect samples of plants and mineral waters.

The results of these expeditions, as well as other scientific news, were published in the *Gazeta de México*, a journal sponsored by King Carlos III and edited by Felipe de Zúñiga y Ontiveros. Upon receiving this commission, de Zúñiga y Ontiveros himself created a questionnaire that was distributed by the viceroy to local officials of New Spain. Among other things, the questionnaire asked for information about "health baths."[23] Perhaps prompted by this questionnaire, on October 22, 1784, Joséph Ignacio Bartolache and Miguel Fernández of the Real Tribunal del Protomedicato traveled eight miles north of Mexico City to examine the Santa Cecilia springs. After bathing, drinking, and examining the waters they found them to "promote and increase urine, cure indigestions, and dissipate 'hypochondriacal gases.'"[24]

The hot springs of the region of Valladolid, what is today Michoacán, were of particular interest to these scientific expeditions. In May 1790 a large group including scientists from Mexico's Royal Botanical Garden (Real Jardín Botánico), illustrators, servants, and Indians left Mexico City leading a mule train laden down with gear such as compasses, thermometers, glassware, and chemicals. This, the third *salida* of the Royal Botanical Expedition, had an itinerary that included the provinces of Michoacán and Sonora to the west and north, and the task of recording information about the geography, botany, and other resources of these regions. The group reached the capital of Valladolid (now Morelia) in August and was received warmly by Governor Riaño and the captain of the royal troops, Joséph Bernardo de Fonserrada. As we have seen, Riaño was a naturalist with an interest in analyzing hot spring waters and identifying their medicinal properties and applications. In the morning he escorted the group and their retinue on a field trip to the nearby hot spring of Cuincho, located on a hacienda owned by the Augustine religious order two and a half miles northwest of Valladolid. The group inspected the modest bathhouse, with its two rooms, each housing a large tub, and looked over the pipe that brought the hot water from the spring to the bathhouse, as well as another nearby spring that issued cold water. The officials explained the different qualities and uses of the waters to the visiting scientists.[25]

Then the scientists got to work analyzing the waters. They examined the springs, as well as the land and plants around the springs, and took measurements. They water was hot—24 degrees on the Reaumur thermometer (an alcohol-based instrument that divided the range between freezing and boiling into 80 degrees)— and it had the weight of distilled water when measured by Beaumé's areometer (a hydrometer used for determining the specific gravity of liquids that were lighter than water). They also looked at, touched, tasted, and smelled the water, finding it to be odorless and colorless, with a flavor of acid. To determine the mineral contents of the water the scientists stirred lime (probably CaO: calcium oxide) into two liters of water, which resulted in a precipitate of twenty-three grains of carbonic acid, indicating that there was carbon dioxide (CO_2) mixed into the water (H_2O).[26] Other reactants produced no precipitates, and so the group extracted the neutral salts by boiling the water and put these aside for analysis back in Mexico City by the director of the Royal Botanical Garden, Vicente Cervantes. They also collected plant specimens around the springs, ordering that information into Linnean categories.

According to the scientists and their guides, the local people frequented the bathhouse to enjoy the "*delicias*" of the hot water and to treat ailments. This dual function shows that the efforts by church and state discussed in chapters 2 and 3 to reshape bathing as a completely therapeutic, health-oriented practice were not entirely successful. The scientists did not comment on the moral or cultural dimensions of bathing for pleasure, which we have seen in the case of Cuincho to have been recently tinged with sexual scandal involving young women and church officials. Instead, they stuck to their chemistry experiments and had quite a lot to say about the virtues of the waters—their medicinal qualities and therapeutic uses. The locals had gotten it wrong, the scientists wrote, mistaking the accumulations of *tequesquite* (mostly table salt and bicarbonate of soda) on the walls of the bathhouse for "nitro," a category of explosive nitrogen-based substances. "Everyone lived with the knowledge that there was nitro in the water, which is a very strange substance to be found in mineral waters," they commented.[27] Under this assumption, and often with the recommendation of a doctor, people bathed in this water with the goal of "tempering the heat in their blood," a condition defined in terms of humoral theories. The scientists declared this faulty analysis and treatment to have led to a failure to cure health problems, and indicated that instead of bathing in it to temper hot blood, people should instead drink it to aid an array of maladies: congested humors, indigestion-producing belches reeking of eggs, "putrid scurvy," and chills.

The chemistry done by these scientists confronted established ideas about the composition of the waters and has been called "the earliest publication found yet concerning Lavoisierian chemistry in New Spain," but their understanding of medicine still did not stray far from the climatological and humoral orientation inherited from the Greeks, Romans, and Arabs.[28] Subsequent studies of mineral springs and their medicinal applications sought more systematic coherence between the modern chemistry employed in testing the waters, and the physiological and chemical understanding of ailments and cures. In 1795 Antonio de Cal, the representative in Puebla for the Royal Botanical Garden and a graduate of its classes on botany, carried out a very similar study of the waters of Tehuacán, which had been used for its therapeutic virtues since the 1600s. And by 1797, in an article in the *Gazeta* discussing the hot springs of Xochitepec, the nomenclature of the new science of chemistry was deployed in the analysis of the spring waters, in the diagnosis of the ailments, and in the explanation for the efficacy of the waters and their mineral contents in alleviating the ailments.[29]

The old ontological assumption that waters had "virtues" survived the emergence of chemistry relatively intact and was even strengthened in some ways. The idea that waters were agential, and that this agency could be described in terms of the effects waters had on other bodies, underwrote the idea that waters contained some chemical substance that could be separated out and utilized to treat some similarly chemical problem in the human body. This concept was also at work in the emerging botanical science practiced in the Royal Garden, which was aimed at discovering and utilizing the useful ingredients in plants. The science of waters practiced in Mexico at the end of the eighteenth century thus marks the transition from the experimental, empirical methods of trial and error central to Galenic medicine, to the theoretical and systemic approach ushered in by Lavoisier. This was a fundamentally new way of understanding water as a universal element containing diverse mineral contaminants, as perceived and measured with increasingly precise techniques. But the scientific view subsumed, rather than abolished, the view that waters were multiple and heterogeneous substances whose virtues derived from their environment. Despite the paradigm shift, then, intellectuals did not relinquish the ontological assumption that waters were virtuous.

PEÑÓN DE LOS BAÑOS

Peñón de los Baños is the name of the small extinct volcanic hill that today lies a few kilometers east of the Zocalo of Mexico City, next to the Benito Juárez international airport. Five hundred years ago, at the moment of contact with Europeans, it was an island in Lake Texcoco, controlled by Aztec emperors and used for hunting. During the colonial period it was an especially important site for the reconfiguration of knowledge about hot springs in New Spain. But more importantly, the history of Peñón in the late eighteenth century reveals how different bathing practices and ontologies of water came together to structure access to the springs. As government reformers and entrepreneurs increasingly set their sights on springs during the Bourbon era, hot springs became sites of struggle between customary bathing practices and new businesses, between rich and poor, and between forms of knowledge about waters. A long, simmering conflict emerged between wealthy clients drawn to the new business of leisure bathing that emerged at that time, and the mass of people who had long used the springs for therapeutic ends.

Peñón de los Baños is one of the best-known hot springs in the history of Mexico, but there is no record that its waters were used for bathing before the Spaniards arrived. The island itself was awarded by the Crown to the conquistador Diego de Ordaz in 1539, and by 1554 there was already a "fine vaulted building . . . a health bath for sick people" who used the waters despite the general distrust the Spaniards had for bathing at that time.[30] Among the users was Fray Alonso Urbaño, a Franciscan monk whose use of the baths to cure his crippled feet and hands prompted a visit in June of 1585 from officials of his order who were worried about immorality among their brethren in the New World.[31] Because his bathing had medical purposes and was backed up with scholarly reasoning (he was described as a "learned and principled" man) he received no sanction. And, as bathing became more accepted by the end of the sixteenth century, the Peñón bathhouse was increasingly frequented by people who were not sick.[32] Despite popular use for pleasure, members of the Spanish religious and political elite continued to justify bathing in the waters of Peñón principally as a medical practice undergirded by Galenic concepts of health. In 1614 Hernando de Deza, inheritor of the bathhouse and spring, turned the facilities into a trust, the property and proceeds from which to be given over to an orphanage.[33] De Deza also commanded that the trust pay for a priest to offer mass in the chapel at the bathhouse complex every day, as well as two "black slaves" to shuttle clients back and forth from Mexico City across Lake Texcoco in canoe. These stipulations were not observed by the descendants of de Deza, who abandoned the bathhouse to disrepair amid legal battles over ownership.

In the eighteenth century the practice of bathing in mineral waters became more widespread and extended through newly emerging social groups, and its purposes and justifications diversified. Peñón, abandoned amid legal battles among the heirs of de Deza over ownership, attracted increasing attention by the illustrious, including doctors, scientists, and officials of the government of Mexico City. In 1755 Antonio Villaseñor y Sánchez wrote that "the owners keep it in a state of neglect and great discomfort, even though the usefulness of its waters are well-known."[34] The doctor Joséph Dumont, newly arrived to Mexico City in 1740, made his first stop the springs at Peñón de los Baños. Medical and chemical research on mineral springs was flourishing in Europe at that time, and doctors often established their practices in a spa town. Dumont immediately located Peñón as a local curing water source, and influenced by the European example he set to examining the waters of Peñón and prescribing them to his patients.

Many people used the springs to treat their ailments, but this use grew in the mid-eighteenth century, backed by a surge in interest among scientists. In 1752

Dumont and his colleague Nicolás de Torres were commissioned by the Real Protomedicato to write a study of the waters of Peñón that would explain their "qualities, virtues and uses" with an eye to rehabilitating the abandoned installations for use by a growing bourgeois clientele. They couched the study in religious, biblical terms, a requirement for any scholarly production in a Hispanic world that was deeply skittish about science and secularism in the context of religious conflicts that mapped onto geopolitical struggles in Europe. Dumont argued for the healing properties of mineral waters using references to the Pool of Bethesda (mentioned in the biblical books of Kings and Isaiah, as well as the Gospel of John) which was renowned for its curing properties. He then moved to a scientific discourse, citing secular authors and texts. Following the climatological ideas of Hippocrates's *Airs, Waters and Places,* he argued that all landscapes produce certain health problems but also provide the elements for their treatment. The humidity of Mexico City, with its lakes and rainfall, caused rheumatism, sciatica, and gout, and God saw fit to provide the waters of Peñón to cure these ailments. Friedrich Hoffman's idea that the body was a hydraulic machine that could be influenced by baths and mineral springs was particularly important to Dumont, but hovered in a strange tension with the mysticism of his biblical references.

Dumont resolved the contradiction between religious and scientific positions by explaining that the healing dynamics of mineral waters on human bodies were designed by God: "[the waters] are a natural pharmacy . . . put there by the powerful, wise and liberal hand of the Divine Architect."[35] Reaching beyond previous prescriptions and biblical references to bathing in mineral waters, Dumont followed his colleagues in Europe by proposing drinking the waters. The curing properties of the waters of Peñón were "acclaimed by the People," Dumont states, but he explained how they work by using the science of chemistry, and providing information about how best to use them. For his part, Nicolás de Torres founded his arguments even more explicitly on chemistry, drawing upon Linnean-style classification schemes for mineral waters based on their "active principles," which produce 479,001,070 possible combinations and thus kinds of waters. He also insisted that clients only use the waters under the care of a professional doctor.

Torres and Dumont were part of a government commission charged with developing the facilities at Peñón to receive and treat an expanding clientele of more wealthy citizens, which, under the influence of European ideas and practices, had a newfound interest in hot springs bathing and the money to pay for it. The inheritors of the original bathhouse were embroiled in litigation over who would control the property and the earnings it generated, and this compelled the government to step in. Along with Torres and Dumont, two architects, a judge, and a scribe made up the party that visited the hot springs in 1752. The facilities had not been renovated since the creation of the De Deza trust in 1614. There was a large bathhouse with three four-room apartments circling a patio, each with a large tub room, and rooms for sleeping. The baths had acquired names and personalities over the centuries—Santa Teresa, el Colorado, la Marquesa—but the buildings were a complete shambles, with crumbling roofs and walls and uneven dirt floors. A family had at some point settled in one of these buildings and looked after the installations. The commission found the baths to be worth very little and unlikely to attract customers, both because of the ruinous state of the buildings and because there was no transportation from Mexico City: the canoes and slaves stipulated by the 1614 charter that created the bathhouse trust never materialized. Backed by the *Informes* of the experts, the royal government forced the transfer of ownership of the bathhouse from the De Deza family to Carlos José Dueñas Pacheco, through the mechanism of an auction that only Dueñas attended. The bathhouse was rebuilt with more space for visitors, and the works were finished in 1765.

Despite the claims of the government and the businessmen who took over the baths, Peñón did indeed attract visitors in its decrepit state and had done so for centuries. These poor people brought their own concepts of curing and were unlikely to afford the cost of a doctor and a new, renovated bathhouse. They appear only as shadows in the archive, as absent causes of events that do not quite add up. The original pretense of the expropriation and renovation of the bathhouse was that there were no visitors because of the poverty of the installations and the costs of transport. However, in an open letter published in the *Gazeta de México* in 1794, the new bathhouse operator Andrés Cabellero wrote that one of the stipulations made by the government for the sale of the hot springs in the 1750s was the construction of a "bath with its corresponding apartment so that poor invalids with no income could make use of its well-known benefits," a requirement that he gladly fulfilled.[36] This flow of poor and indigenous patients to Peñón was augmented after the 1764 reopening when, in 1768, the Hospital Real de Naturales contracted to send its indigenous patients daily for treatments.[37] Caballero was willing, he promised, to open more of the renovated baths to poor people for a quarter of the regular price, if his finances permitted.

There was clearly an established use of the bathhouse among the poor and the indigenous, but the renovation of the bathhouse at Peñón de los Baños was aimed at a new group of relatively wealthy bathers, and in fact the new managers restricted access by poor folks to the resource. In 1791 the city council passed an act prohibiting Peñón's administrator from limiting free access by poor bathers to the bathhouse, or stopping them from taking bottles of water to be used as medicine.[38] Very few of those who had previously frequented the bathhouse with its dirt floors and collapsing roofs could afford the new two-peso/twenty-four-hour rate, nor were they likely to purchase the *pulque,* fine wines, and drinking chocolate on sale at the new store. It was unlikely that the administrator had them in mind when he assured the "Public" that all kinds of the most select food was available at the store, "not just the everyday kinds, but also those that the educated and gracious visitor could entertain guests with."[39] The *plebe* certainly would not make use of the two horse-drawn coaches that brought visitors to Peñón from Mexico City (for two or three pesos), nor was it very likely that they would have enough stuff with them to fill the wagon that was for hire. If bathers came on their own horses, and paid to stable them at the bathhouse, they certainly were among the most economically privileged of the public.

The new Peñón bathhouse offered all the comforts that could be found at the spas that were popping up all over Europe to serve the new leisured bourgeoisie: clean rooms, beds, linens, food, drinks, and socialization with peers. After the renovation, "the baths were heavily visited year-round," and the leisured elite came flocking to the site.[40] In 1777 the archbishop of Puebla, Juan de Viera, recounted that because each apartment had its own bath, "sick people bathe there freely without being registered; usually there are 8–10 families using the bathhouse."[41] But the medicinal benefits were not the only, or even the principal, attraction for many users, and soon it became a preferred destination for well-to-do urbanites looking to relax in the countryside. According to Viera, "because the bathhouse is so large . . . and offers water, greenery, solitude and tranquility, many go there not to bathe but rather to have fun, hunting ducks and rabbits, attending concerts, dances and big banquets that often last two or three days."[42] In 1793 taxis were shuttling clients to Peñón, and by 1797 a businessman had opened a luxury transport service that carried patients from Mexico City to Peñón in litters.[43]

In keeping with the model of the European spa, Caballero asked doctor Gabriel de Ocampo to publish a scientific, medical statement in conjunction with the announcement of the opening of the renovated bathhouse in the *Gazeta de México*. Ocampo aimed at "encouraging patients in need to use the waters," offering the bather an analysis of the therapeutic uses of the waters of Peñón based in the quickly evolving sciences of chemistry and medicine and in particular the study carried out on Peñón's waters by the Royal Botanical Expedition in May of 1790.[44] High temperatures and dissolved gas opened the pores of the skin to allow substances to enter, and the salts, gas, and heat combined to "unblock, relax and dissolve, without taking away the necessary tone of the body's fibers." These properties encouraged the circulation of liquids, unblocked the nervous system, helped with gout and rheumatism, constipation, digestion, and appetite. The exact mixture and proportion of gases, salts, and heat in Peñón's waters was a product of the "sovereign and supreme architect"—God—and could never be replicated in the doctor's office. So only by using the new bathhouse could the "virtues" of the waters be obtained, and only then under the care of "a circumspect and trained doctor." The waters, he warned, could have very negative effects if used improperly. Treatments should be done for ten or more days to gain the "almost miraculous" results, and they could be carried out on people from less than ten years of age to more than eighty.

The unequivocally medical justification for the existence and use of the Peñón bathhouse that is found in Ocampo's letter strikes an uneasy counterpoint with the assurances by Caballero that the bathhouse would provide all the bourgeois comforts, and with the descriptions elsewhere of three-day festivities held there in the 1760s. This is the balance that bathing represented for Mexico's elite since the sixteenth century—of promise and peril, rulers and ruled—only resignified in the Enlightenment discourses of science and citizenship that bubble up in the Spanish empire during the late eighteenth century. Clearly this was a balance that governments in Europe were also negotiating, but in Europe the impact of the bourgeois revolution on water cultures seems to have been steadier, deeper, and more lineal. In Mexico, on the other hand, the emergence of scientific bathing and the policing of water was contested in ways that are not visible in the literature on European spas and hot springs. Humble bathers continued to gain access, one way or another, to the waters, and continued to use them in accordance with ideas and beliefs passed down through generations. And, as we shall see, these long-standing bathing traditions persisted alongside bourgeois involvement in the business and science of bathing in Mexico.

CONCLUSION: THE CONTRADICTIONS OF ENLIGHTENMENT BATHING

Bathing changed in the eighteenth century. Ideas about health and water rooted in the long tradition of humoral and climatic thought passed down from antiquity through medieval Christian and Arab scholars were reworked by scholars who took a scientific experimental and conceptual approach. Empirical knowledge about the benefits of different waters was systematized according to new categories and classifications generated by chemists. While scholars in Iberia and the Americas continued to frame the study of the physical and natural world in religious terms, this took second place to the descriptions of elements and physical relationships themselves, and how the stuff of the animal, vegetable, and mineral kingdoms influenced wellness and illness in people's bodies. God was given credit as the supreme author or architect of these patterned relationships, but did not much occupy the attention of the new scientists. Nor was religion central to the efforts of government to rule well. The scientific approach to governing was conceived more as a project of managing territories, environments, and populations to ensure order and well-being than in moral terms. Ideas, beliefs, and practices were still central to policing, but increasingly they were disassociated from notions of right and wrong based in spiritual authority. Instead, ideas of civic order and public health guided the proclamations and efforts of government to rationally manage the human relationship to water. More than sin, perhaps, the peril of social bathing was disorder.

The Enlightenment reframing of water culture was rooted in a deep, telluric transformation of economy and society in the eighteenth century that generated new social groups and activities. The growing bourgeoisie in Europe had disposable income and leisure time, and this opened up possibilities for making a business of bathing. Spas offered therapy, rural tranquility, and social settings for newly wealthy urbanites. Capital, loosened from moral and political limits of the old order, expanded and flowed toward investment opportunities such as medicine, bathing, and infrastructure development. Although Spain and its colonies participated partially, unevenly, and somewhat belatedly in this historical emergence of capitalism, the Bourbon governments embraced the role of promoting these trends in the water sector by supporting expeditions to study hot springs, and the renovation of bathhouses and urban water systems.

The business of bourgeois bathing was not very successful in New Spain. The Peñón bathhouse was located a few kilometers from the wealthy inhabitants and cosmopolitan tastes of the capital, which made that particular site attractive to some investors. But spa promotors confronted deeply rooted peasant and plebian water cultures based on social and economic principles that were at odds with commoditization. Open access and communal property were prickly obstacles to the enclosure of hot springs, and Peñón's development as a business moved forward only with the concession that at least one bath would remain open to the poor who traditionally used those waters. Hot springs such as San Bartolomé, Topo Chico, or those in Aguascalientes, far from the capital and its emergent bourgeoisie, would remain undeveloped until improved transportation infrastructure enabled spa tourism in the late nineteenth and twentieth centuries (see chapters 7 and 8).

Even with its central location, the bourgeois bathhouse at Peñón soon lost momentum during the War of Independence (1810–21). In 1827, a few years after the wave of nationalist mobilization forced the Spanish Crown to cede control over Mexico and most of its American possessions, José María Manero wrote a letter to the government of the City of Mexico complaining that all the gains made during the rule of Revillagigedo in ordering, rebuilding, and controlling bathing practices at Peñón de los Baños had been lost. The owner of the bathhouse had abandoned his obligations to "repair and maintain the guest rooms, provide all the necessary goods [soap, sponge, towel, etc.]; to keep them clean."[45]

Manero couched his argument in terms of the public need for and benefit from the baths, thus positing the existence of a national citizenry and of public health as a domain of political and economic intervention. What his complaint to the city government did not stress is that people of humble origin had used Peñón's springs and installations before the bourgeois rebuilding, that they succeeded in maintaining some access to it when it was turned into a business, and continued to make use of the facilities when wealthier Mexicans shied away from their decrepit condition. Manero himself saw "many poor sick people" continue to go there to cure their ailments, as did patients under treatment by the hospitals. This use was year-round, and responded to the unpredictable appearance of ailments rather than that of the summer spa season that defined elite bathing in Europe. So in fact much of the "public" was indeed using the bathhouse. But like most of those scholars and officials whose words are recorded in the journals and archives, Manero used the universal category of public to reflect the position of a far smaller, elite class of people—the rulers, not the ruled.

Humble folks used the hot springs and bathhouses of Mexico throughout the early nineteenth century, as popular water cultures resumed their ancient courses and the Enlightenment visions of a well-policed bath faded. That project of the bourgeois spa, with its genteel leisure practices and scientific purpose, remained strongly alluring for the elite, however, especially for those who had contact with bathing culture in Europe and the United States. The business of bathing was in latency, awaiting the next cycle of accumulation to mobilize the building of infrastructure, the florescence of concepts, and the policing of practices.

"Like most great ideas of Spanish days, it is now in a state of perfect desolation, though people still flock there for various complaints. When one goes there to bathe, it is necessary to carry a mattress, to lie down on when you leave the bath, linen, a bottle of cold water, of which there is not a drop in the place, and which is particularly necessary for an invalid in case of faintness—in short, everything that you may require. . . . We could not help thinking, were these baths in the hands of some enterprising and speculative Yankee, what a fortune he would make; how he would build a hotel a la Saratoga, would paper the rooms, and otherwise beautify this uncouth temple of boiling water."

—Fanny Calderón de la Barca, 1843

Source: Calderón de la Barca 1843: 403–5.

Groundwater and Hydraulic Opulence in the Late Nineteenth Century

"Well, in our country," said Alice, still panting a little, "you'd generally get to somewhere else—if you ran very fast for a long time, as we've been doing." "A slow sort of country!" said the Queen. "Now, here, you see, it takes all the running you can do, to keep in the same place. If you want to get somewhere else, you must run at least twice as fast as that!"

—LEWIS CARROLL, *THROUGH THE LOOKING GLASS*

In the second half of the nineteenth century artesian wells tapped into groundwater, ending centuries of water scarcity and greatly expanding access to baths in Mexico City. The individualized immersion bath (*placer*) once offered to wealthier, more European clients was now available to almost everybody. Many of the downtown bathhouses that served humble city dwellers shuttered their *temazcales* and replaced them with low-cost wooden *placeres* grouped together in a shared room.[1] These humble bathhouses charged for each bucket of hot water, but usually provided all the cold water a client wished, and they used much more water for their wooden *placeres* than they had for the *temazcales* that preceded them. At the same time as the *placer* was being adopted by the masses in the old bathhouses of the city center, new and exclusive bathing facilities were sprouting up on the western side of the city along the Paseo de la Reforma that offered both social and individual contacts with great volumes of water in a variety of forms including swimming pools, tubs, steam rooms, and showers.[2] Bathers in both new and old bathhouses luxuriated in an unprecedented hydraulic opulence provided by seemingly unlimited groundwater from artesian wells.

The abundance of groundwater reshaped the practices and social relations of bathing in Mexico. In this chapter I discuss how, around 1850, bountiful, clean water was supplied in places that had not been served previously by existing infrastructure, and in quantities that enabled bathing with more frequency, with more water. Existing bathhouses turned into immersion baths, and many new lavish, modern bathing centers, or *balnearios*, were built that offered a much wider range of contacts with waters, including swimming, diving, wet and dry saunas, drinking fountains, and splashing pools for children. Groundwater filled pools in urban and rural settings, expanding the practice of swimming for fun and fitness that was before mostly limited by access to natural bodies of surface water. The expansion of bathing in the late nineteenth century was backed by a new assumption that water was available in large amounts—a structured feeling of hydraulic opulence that emerged along with artesian wells. The political ecology of groundwater presented in this chapter shows how infrastructure, bathing practices, and concepts of cleanliness evolved together.[3]

Easy groundwater during late nineteenth- and twentieth-century Mexico made both social and individual bathing more common. Jeff Wiltse shows that in the United States after 1940, extensive water infrastructure enabled the proliferation of private backyard pools, resulting in Americans "bathing alone" rather than together in public pools.[4] In the late nineteenth and early twentieth centuries in Mexico we see a similar growth of the water supply, but the individualization of bathing was a more ambivalent process. Despite the availability of water, and the connotations of luxury and modernity offered by private bathing, Mexicans continued to bathe together. Poor people soaked and scrubbed in individual *placeres* in large shared rooms, and wealthy Mexicans met to socialize and take the waters in elaborate bathhouses. While the hygienic and sanitary function of bathhouses slowly moved to household bathrooms in the twentieth century, social bathing for fun and fitness continued to flourish in the country's baths.

THE GROUNDWATER REVOLUTION

The scarcity of clean water that characterized the colonial period continued unabated after independence, as did government efforts to police the shortage through the identification of new sources and the construction of infrastructure. As was discussed in chapter 4, for centuries water from the aqueducts of Mexico City was concessioned to wealthy property owners or delivered to the public fountains. There were also shallow wells used by the city government for cleaning the streets, and the plebian mass often used these for their houses, their animals, themselves, and their clothes, as they did the water that flowed through the city's drainage canals and the rivers on the outskirts of the city. People liked to wash and bathe in the water of the wells and drainage canals because it was softer than the city water and produced more suds. More importantly, it was available and free. But the quality of those waters was dubious and the public bathing they supported was frowned upon by many.

In the 1830s city officials sought to increase the supply of water, and set their sights on the Xacopinca spring, located to the north near the towns of Azcapotzalco and Tacuba. This was a spring that first served the pre-Hispanic settlements on the islands in the lake, conducting water through an aqueduct to the town of Tlatelolco, north of Tenochtitlán. In the 1400s this water system was overshadowed by the aqueduct built to carry water from the Chapultepec springs to Tenochtitlán. After it was destroyed by the Spaniards, the Xacopinca aqueduct was only restored to working order for a brief time at the beginning of the seventeenth century, despite periodic renewed interest. The city owned the spring, and in 1839 once again studied the possibility of integrating Xacopinca into the city's infrastructure, calculating that the sale of the water could pay for the works. In 1843 the city signed a deal with a private investor who offered to finance and build an aqueduct from the Xacopinca spring to the fountain in the plaza of Tlatelolco in exchange for the right to sell the water. In order to stop the existing access and uses of the water by local peasants, the business was given permission to build walls around the water source.

The chemist Leopoldo Río de la Loza was commissioned to conduct an analysis of the mineral and biological contents of the Xacopinca waters.[5] He found it to be "better" (less dissolved solids) than the *agua gorda* (hard water—literally "thick water") of Chapultpec and "worse" than the *agua delgada* (soft water—literally "thin water") of Santa Fe, and concluded that channeling the spring would have the double benefit of providing potable water for the city and removing an unhealthy swamp at the site of the spring.[6] The gradient for the kilometer-long aqueduct to deliver the water to Tlatelolco was adequate, but by the 1850s the springflow had dwindled so much that there was not enough pressure to move the water down the aqueduct, and the quantity of water was insufficient to justify the expense.[7] With this plan to increase supplies of water frustrated, in 1854 the city administrators instead reduced the amount of water delivered to each user by installing "economical faucets" throughout the system, a measure that raised the ire of the bathhouse proprietors.[8]

Despite the scarcity of water, the demand for baths continued to increase, and businessmen built new baths or expanded existing ones. Santiago Vega founded the Baños del Amor de Dios in 1853, and in 1866 asked for a fifty percent increase in the water concession. In 1856 José Guadalupe Velásquez asked for "one more concession" of water for his *baños* at Number 11 Calle Don Toribio, and the next year Manuel Murguía petitioned the city for a *merced* of five "pajas de agua" (about 10 cc/second) for a new bathhouse he aimed to build in the Plazuela de Juan Carbonero. The bathhouse was built and the water delivered, but subsequent pleas for more water in 1861 and 1877 suggest that the number of people in the city who wished to bathe kept growing.[9]

After 1850, the perforation of artesian wells seemed to erase the limits to the supply of good quality water. Engineers armed with new drilling equipment and techniques opened hundreds of artesian wells in the Valley of Mexico during the last half of the nineteenth century, part of a global groundwater revolution.[10] Artesian waters emerge under their own pressure, without pumping. Wells are drilled in a location where the surface of the ground, or wellhead, is lower than a portion of the aquifer that lies above the wellhead, and so the water flows downhill within the aquifer and then flows up and out of the well bore. This is common for aquifers located in sloping land, such as the Valley of Mexico. Artesian wells mimic naturally occurring springs, the difference being that a route is opened artificially by a drill for the water to reach the surface. The Valley of Mexico was a geological formation that was suited to artesian wells, as there were altitude differentials in the subsoil water that created hydrostatic pressure that was maintained by geological formations of alluvial silt and clay. In that context, water would spring unaided anywhere that a well opened access to that confined aquifer.

In Mexico, attempts had been made in the early part of the nineteenth century, but artesian wells only became common in the 1850s.[11] This was largely due to the efforts of Sebastián Pane, who by 1854 had opened at least 20 artesian wells, and by 1857 had completed 144 wells, 24 for use in irrigation and the control of dust on public roads, and 120 for the houses of individuals.[12] Pane used the "Chinese" system of drilling, a technique of percussion drilling with a heavy chisel on the end of a rope, which was pioneered a thousand years ago to tap water and natural gas with wells hundreds of feet deep in the province of Szechuan.[13] Soon others were using a different system that enabled even deeper exploration, and by the 1860s there was a lively discussion of drilling techniques and many companies were operating drilling rigs in Mexico.[14] Pane, however, continued to lead the industry, completing hundreds of artesian wells in Mexico City, as well as in the cities of Veracruz, Tampico, Cordoba, Manzanillo, and Mazatlán.[15] In the 1850s he opened a business office on a plot of land on Paseo de la Reforma, where he contracted to build wells for individual houses or groups of houses, and a few years later he built the famous Alberca Pane bathhouse on that property.[16] He even received permission from the Ministry of Development to experiment with using wells to desiccate Lake Texcoco by draining its waters into the subsoil.[17]

The advent of the artesian wells brought easy water, and a hope of finally resolving centuries of water scarcity and the social struggles and policing engendered by it. Most of the wells sunk in the city by Pane in the 1850s were for "private houses," but three of them—at Los Migueles, Bucareli, and Cordobanes—provided water to the public in the city center to supplement the Santa Fe aqueduct and the springs of Chapultepec.[18] In 1863 Pane signed a contract with the city's Comisión de Aguas to open eight new artesian wells in different plazas in the historical center of Mexico City, and in 1869 the city ordered another three wells drilled for neighborhoods that did not have adequate water service.[19] Some, such as the well in the plaza of Salto del Agua, served existing public fountains. In 1871 two more were sunk near San Lázaro and in 1872 twelve more public wells were drilled, with five of those to the west of the city center on the Paseo de la Reforma.[20] Still, most wells were private, and served the wealthy. By 1883 Mexico City had 483 wells, thanks to a growing professional cadre of engineers with drilling equipment.[21] About a third of these wells were located in the city's Octava Demarcación, which included the new, wealthy neighborhoods to the west of the city center along Paseo de la

Reforma, from Bucareli to Chapultepec.[22] The bathhouses and swimming pools that were all the rage in the era of President Porfirio Díaz (1877–1911, known as the "Porfiriato") were located in this area, and they were served by artesian wells. The flurry of drilling led by Pane almost doubled the entire water supply in the city by the 1860s and almost tripled it by 1883.[23] In 1895 public artesian wells provided about a quarter of the water that coursed through the city's distribution system.[24]

Hundreds of artesian wells were not capped, but rather left to flow freely and create wetlands that city leaders viewed as dangerous to public health. Near the Chapultepec springs, "three or more artesian wells without faucets" coursed, puddled, and eventually mixed with waters leaking from the aqueduct that supplied the fountain at Salto del Agua.[25] But even without this wastage, the huge increase in water supply between 1853 and 1883 saturated the drainage system in the city, adding to fears of "infectious waters," miasmas, putrefaction, and other threats. In response to this new abundance of groundwater, the city government issued a series of dispositions regulating extraction but, more than anything, disposal of the liquid.[26] In some cases the government required well operators to channel their excess waters to public fountains, but these regulations were generally not enforced.

There were signs from the beginning of the boom that groundwater was finite, but not everyone wished to recognize them. Wells began to dry up and dwindle just a few years after they were sunk.[27] When Antonio Peñafiel visited the Xacopinca spring in the 1880s to assess the hygienic qualities of its waters, he noted that the water pressure had fallen so much that the spring was stagnant, although he failed to link this fact to the rise in the number of artesian wells he documented elsewhere.[28] The profusion of new artesian wells also caused water pressure to drop significantly in the aquifer that gave rise to the Chapultepec springs, reducing the flow through the aqueduct to the fountain at Salto del Agua. But when asked by the city about the likelihood that these shortages were caused by the installation of an artesian well nearby in the Hormiga parcel (now next to the presidential residence in the Bosque de Chapultepec), engineer Francisco Herrera rejected the possibility, arguing that the springs were just clogged.[29] Others in the city government pointed to the perforation of artesian wells, but this argument did not gain traction among scientists such as Leopoldo Río de la Loza and Ernesto Craveri, who dismissed the "common doubts of those who fear that these waters are not permanent," saying that it would be "very strange" if the main aquifer (the third one down) lacked water, even in the dry season.[30] When the flow of waters from many artesian wells dwindled in the 1870s and 1880s, the usual diagnosis was not a reduction of water pressure due to overextraction, but rather that the wells were clogged.[31]

Unlike the intensely managed and legislated surface water, it was not at all clear how to manage or resolve conflicts over groundwater. For example, a well was opened in the 1860s in the public plaza of Atzcapotzalco that served a group of nearby homeowners.[32] When another well was drilled nearby by another homeowner, this first group lodged a legal case, claiming that the new well reduced the flow of water to their previously existing one. Because there were no legal prec-edents for determining the property status of, and rights to, groundwater, Río de la Loza was asked to comment on the matter. He suggested that the new user should utilize a different aquifer, at a different depth, but was the first to admit this was not a solution that could be made universal.

Instead of confronting conflicts over groundwater by regulating extraction, the city government sought once again to increase supply. Reduced flow from the Chapultepec springs prompted a two-pronged effort. First, the government purchased titles to *mercedes* to increase the amount of water it owned in the Albercas de Chapultepec. The Alberca Chica was fed by the springs and supplied the Belen Aqueduct. The Alberca Grande, also known as the Alberca Exterior, was a deposit on the southern edge of the Bosque (where Avenida Constituyentes is today) that was used to supply a house, fields, and orchard in Tacubaya owned by José Amor y Escandón, a descendent of the Conde de Miravalle.[33] The title to this water was purchased by the city, along with titles to the "vertientes del Bosque de Chapultepec," which were waters that flowed from the Chapultepec springs, but which were not captured by the Alberca Chica and the aqueduct and flowed southward out of the Bosque.[34] There was another reservoir, farther along, belong-ing to the Hacienda de la Teja.[35] In addition to these water purchases, the city gov-ernment cleaned and fixed the Alberca Chica so that it would store more water. As a result of these measures, in 1870 the engineer Manuel Patiño reported that the leaks in the Alberca Chica were all repaired and soon there would be more liquid than could fit in the aqueduct.[36]

This remedy for the shortage of water for the city came at the expense of those who, like the large landowner José Amor y Escandón or the inhabitants of the barrio of San Miguel Chapultepec just south of the Bosque, had used these waters previously.[37] At the end of the nineteenth century this area lay on the outskirts of the city, and Amor y Escandón owned a swimming pool with dressing rooms that was a popular destination for both city dwellers and foreign travelers such as Gilbert Haven, who described an artesian spring that was "the private prop-erty of Señor Escandón, who makes many a penny out of its waters."[38] In a letter explaining his grievances, Amor y Escandón argued that the city's effort to rem-edy the water scarcity led it to take "all the measures necessary" to capture the flow from the hardwater springs of the southern part of the Bosque, including building levees and dikes that prevented the water from reaching his *alberca*, or the lands of the Teja and Condesa haciendas, to whom that water had customar-ily belonged. Amor y Escandón told the city that the loss of water forced him to close his *baños*, countering the assumption that the city water supply should be the foremost priority with the argument that the baths were beneficial for pub-lic health.[39] It is ironic that at least some of the water that was taken from the Escandón swimming pool in fact ended up in the city's bathhouses after a long trip through the aqueduct.

"We pass out of the gate [of Chapultepec Park], ride under the shading willows by the watercourses, enter the gardens of the bath, and the enclosure of the spring. Here is a pool fifty feet square and forty feet deep. The water is so clear that you can see it breaking out of the rock-bed . . . amidst the ferns and grasses that cover that natural floor with a perpetual carpet. Here to plunge you will find delightful. . . . An adjoining square the water flows into, whose floor is paved with tiles, and whose depth is not above your neck. . . . A like bath for ladies is nearby, and a saunter in the garden follows the refreshment."

—Gilbert Haven, 1875

Source: Haven 1875: 224–25.

As wells dried up, even more were perforated. In 1900 the city counted 1,200 artesian wells, and they were common outside of Mexico City as well. Jalisco, San Luis Potosí, Querétaro, and other states had wells for irrigating cities and haciendas by the 1860s. In Celaya, Guanajuato, artesian wells tapped thermal aquifers to supply bathhouses and water fields.[40] In 1873, a traveler to that city described a "bathing establishment, which is supplied by warm water from an Artesian well . . . There are a series of private compartments, and a large public basin sufficiently deep for swimming purposes."[41]

The access to groundwater between 1850 and 1900 revolutionized water culture in Mexico, marking the beginning of an age of hydraulic opulence and optimism that continued through much of the twentieth century. Consumers, in particular, quickly grew accustomed to this new supply of water and practices such as frequent bathing by immersion and swimming that the water enabled. Some water managers and scientists also assumed that the centuries-old limits to Mexico City's water supply had been overcome, despite evidence that subsoil water was finite and its extraction caused environmental health hazards and land subsidence. However ephemeral it was, plentiful cheap water restructured feelings about the relationship of people to their waters and set water managers on a path of increasing supply from which they have not since wavered. Aquifer depletion reduced hydrostatic pressure and many artesian wells stopped flowing, but the advent of electrical service at the end of the nineteenth century made it easy to pump the water from those wells. People got used to having copious quantities of fresh potable water at hand, and this assumption of hydraulic opulence drove ever-greater efforts to harness water. To sustain the expansion of water consumption and supply, in 1884 Antonio Peñafiel set his sights on a bigger prize: the springwaters of Xochilmilco, twenty-five kilometers away on the southern edge of the Valley of Mexico. These works were completed in 1908, the first of many such projects that steadily increased the water consumption of Mexico City over the last century.

THE EXPANSION OF SOCIAL BATHING

Zopopan, Jalisco, Mexico, 1840. A typical Sunday morning in late spring in western Mexico; the land thirsting for the summer rains soon to come. On the road leading out of Guadalajara to the west carriages of the well-heeled jostle past folks walking in *huaraches* [sandals], all making their way to the public baths near the towns of Zoquipa and Atemaxac, a mile and half from the city. The small river is low in the dry season, and there are bathhouses installed all along its course: improvised structures made of carrizo reeds with grass walls and roofs that afford some privacy to the modest and the decent. Men and children sit and splash in the water, laughing and playing, some relaxing. Others have retired to the grassy embankments to enjoy picnics and purchase "exquisite watermelons and sweet melons from Caxititlan" from throngs of itinerant vendors announcing their goods in crisp shouts. Some add to the bustle by singing along with the musicians wandering among the crowds. The young women walk in pairs and threes along the riverbank in their weekend attire, freshly picked wildflowers tucked into hair and hatbands. They are waiting for the waters to warm up so they can take a dip in the early afternoon. Local indigenous people organize the whole affair, renting out the bathhouses, selling food, digging out gravel to form swimming pools in the river, and tending the grassy embankment. The dark clouds assembling overhead in the late afternoon signal the end is near for the short summer season of outdoor bathing. When the rains finally come in force, the locals will take down the bathhouses and store the materials, before the rising waters wash the baths into memory. But come next summer, fun and fashion will once again get urbanites from Guadalajara up early in the morning to make the trip to the baths of Zopopan.

—Ignacio Cumplido, 1842

Source: Cumplido 1842: n.p.

How did Mexicans bathe in the nineteenth century? How did bathing practices change after 1850 with the hydraulic opulence created by artesian wells? How was ecology and infrastructure related to culture? Bathing and swimming, like so many quotidian experiences and activities, are often hard to discern in the archival record. Almanacs and travel literature, however, afford ethnographic glimpses of these activities in Mexico in the nineteenth century. At the same time, these are always partial, selective visions that tell us just as much about the cultural assumptions of their narrators.

In 1842 Ignacio Cumplido portrayed the leisurely bathing practices of elite urban Mexicans in a landscape specially managed for this activity. Alexander Forbes, traveling through Tepic in 1849 and 1850, noted a similar arrangement of bathhouses "made of wattles, and thatched . . . situated at the river side, where the stream is tolerable deep . . . divided into different compartments, and much used by the better class of inhabitants."[42] Bathing among well-to-do Mexicans was commonplace, just as it was for their European counterparts, who during the nineteenth century turned freshwater, mineral water, and seawater spas and resorts into mass leisure destinations. The seaside town of "San Blas," Forbes continued, "is frequented by the Tepiqueños, who during the latter part of the dry season, come down for the sake of sea-bathing."[43]

Poor Mexicans ran the bathhouses for the wealthy, but they too bathed. In 1828 George Lyon visited the Pánuco River near Tampico, and described people "bathing in the river whole families at a time, which appears to be their morning and evening custom."[44] "Such families as choose to," he continued, "devote a little trouble and expense to decency, small spaces are staked off near the banks, and lightly covered with palm branches: but such niceties are not much attended to; both sexes bathe without scruple at the same time, and many of the young women swim extremely well."[45] Steaming up the Río Bravo/Grande in the 1840s, Corydon Donnavan spied "droves of joyous young girls disporting like mermaids among the waters."[46] A decade later on the same river, Emmanuel Domenech noted that "a number of people of every age and each sex were bathing."[47] Obviously impressed by the propensity of Mexicans to take to the waters, another traveler in the 1840s declared, "whenever I was in sight of the river, or the canal of the mills, I could behold men, women and children floundering in the water."[48]

It is clear in these accounts from the mid-nineteenth century that bathing and swimming in springs, rivers, and seas were social activities that were fun for people from all walks of life. In the torrid river drainages of the Gulf Coast bathing and swimming allowed them to cool off and play. But bathing was also about cleaning bodies and clothes. Except for the wealthy, who had servants to do their work, people swam, bathed, and washed clothes at the same time. Alexander Forbes described elite social bathing, but also noted that the humble people in Tepic went to "*pozos* (swimming holes) . . . much used by laundresses and bathers . . . usually large holes dug just below the springs."[49] On the river in the nearby city of Colima, John Lewis Geiger described the "numerous baths erected along its course, and the temporary laundry establishments," all grouped together.[50]

Women, playing and bathing, attracted the interest of travelers, men especially, who never failed to comment on their nakedness. When George Ruxton rode into Querétaro on horseback in the mid-1840s, he was surprised to come across "a bevy of women and girls 'in the garb of Eve' and in open day, tumbling and splashing in the water." When his group stopped to watch "they were attacked by the swarthy naiads with laughing and splashing, and shouts of '¡ay que sin verguenzas!'—what shameful rogues!—¡echales muchachas!—at them, girls; splash the rascals—and into our faces came showers of water, until, drenched to the skin, we were glad to beat a retreat."[51] In Tepic's river Alexander Forbes always found a way to gaze upon "two or three damsels sittin' in it, rather in undress, washing their hair." "The women," he continued, "all have beautiful hair, and seem to take great pains in washing and cultivating it, as you may see them all day in the river by scores."⁵² These passages resound with the orientalist titillation of the odalisque, but nevertheless show that undressing and bathing in public was an everyday and rather unexceptional activity. The forms of modesty, honor, and respect that without a doubt regulated this activity were beyond the comprehension of the travelers, who mistook their ignorance of local scruples for an absence of scruples.

At the end of the nineteenth century the influx of capital to Mexico led to the expansion of infrastructure, economic growth, and the accentuation of class dynamics from the industrial era. These changes had important effects on bathing. When Albert Gilliam visited the Aguascalientes hot springs in 1846 they were "not covered by houses, or shelter of any kind, and both rural poor and city dwellers used them in their rustic state, often seeking cures."⁵³ By the 1880s, however there were "extensive and commodious bathing houses . . . surrounded with flower gardens" for the wealthy in Aguascalientes, but even more so for the new waves of tourists arriving by train from Mexico City and the United States.⁵⁴ While these bathhouses had shared public swimming spaces, their clients washed in individual bathing rooms, where privacy enabled modesty. Those unable to afford such luxuries continued to bathe as always, in the community waters, often in mixed company.

Travelers to Mexico in the late nineteenth century were likely to be adventurous elements of the leisured bourgeoisie from Europe and the United States, riding the new railroads. They were quite aware of the distinctions between wealthy and poor in Mexico, and how these played out in bathing practices. Privacy and nudity were directly connected to social class, and some travelers recognized this. Despite the flood of capital into Mexico during the Porfiriato, and the rapid growth of economy and infrastructure, many Mexicans were so poor that they had only one change of clothes. Blake and Sullivan remarked "nine times out of ten the one suit, noonday and night, forms the entire stock of wearing apparel."⁵⁵ In other words, when women washed clothes, their owners were naked. As Francis Smith put it, there were two classes in the hot springs in Aguascalientes, "those who have something on and those who have nothing."⁵⁶ Cora Crawford, for example, commented that in Aguascalientes there "runs an acequia where all the washer-women of the town gather," and where "the poor congregate because the luxuries of the private bath-houses are beyond their reach."⁵⁷

Travelers in the mid-nineteenth century depicted nude bathing in public as somewhat humorous, but by the late nineteenth century they responded to it as a moral or social problem. On the one hand, consider the shock expressed by Crawford upon witnessing an "inhuman scene" in the hot springs town of Aguascalientes, where, "*en cueros* [naked], and with utter abandon, men, women and children plunge together into the water."⁵⁸ Julia Jackson described the same scene as "startling": hundreds of people "bathing and disporting themselves in the water, with an absence of bathing dresses and an unconsciousness of being visible to the naked eye."[59] On the other hand, some travelers adopted a position of cultural relativism and self-reflective detachment. Frank Collins Baker viewed a scene of men and women bathing naked—each group on opposing banks of the river—with the scientific eye of a naturalist. "It seemed rather strange to us," he reflected, "but was the custom, and of course, aroused no curiosity among the inhabitants."[60] Francis Smith also saw mixed-sex social bathing by the poor as "one of the customs of the country."[61] Others recognized the practicality of this kind of bathing. "Judging from the number of primitive bathing and washing establishments we met by country brooks and city ditches," Mary Blake and Margaret Sullivan wrote, "wherein father, mother, children and clothes were all being cleaned together, I am inclined to think they prefer the public demonstration. And why should they not, if it be simple and easier?"[62] Regardless of the stance taken by the traveler, these accounts reflect a widening gap in the bathing infrastructures and practices of the wealthy and poor in many places in Mexico during the second part of the nineteenth century.

The narratives about bathing display the perspective of the narrators, but they also provide ethnographic glimpses of everyday interactions that people had with water. For example, even as they eroticized Mexican women, travelers saw them as workers, washing clothes and raising children. Donnavan argued that "women perform this very necessary part of household labor, in the river,"[63] and Alsden Case described:

> scores of women contentedly scrubbing, sudsing, and wringing their clothes, for the river bank is the universal washing place in Mexico. Flat stones served as washboards. All the bushes and rocks of the vicinity were decorated with clothes. And the bathers! Children, children everywhere, with skins all shades of brown. Every woman had brought her tribe, and judging from the appearances, their clothes were being washed "while they waited!"[64]

The same scene was repeated in the urban bathhouses. Antonio García Cubas revealed his anxieties about class and contagion when he described the "dirty custom" of mothers bathing their children in their leftover bathwater in the small wooden tubs that lined the collective bathing rooms of the downtown bathhouses that served the poor, rooms still known in 1900 as *temazcales* despite the fact that those steambaths had been mostly eliminated by that time in Mexico City.[65] The gendered division of labor charged poor women with the responsibility for washing clothes and bathing children—their own as well as those of others.

Travelers understood their social distance from and uneasiness with poor Mexicans in the idiom of cleanliness and hygiene. Blake and Sullivan had been told that Mexicans were dirty, but they discovered that bathing was directly related to the availability of water. "We found them dirty," they wrote, "as regards personal cleanliness, in towns like Chihuahua and Zacatecas, where water has to be dipped with a gourd from the basin of a stone fountain, with scores awaiting their turn, or bought from a carrier. But in Aguas Calientes . . . there was no suspicion of uncleanliness."[66] Sometimes a traveler would go to lengths to reconcile the preconception that Mexicans were dirty with the empirical data that they were constantly bathing. "Though every shop in every city keeps and sells vast quantities of soap," wrote Frederick Ober, "and though everybody in the neighborhood of a stream is constantly washing, both himself and his garments, yet [sic] every person of the lower order is as dirty as though just dipped in a city sewer."[67] Others racialized the perception of uncleanliness, assuming that all who did not bathe were Indians. "The bulk of the poorer population of the capital are Indians, who greatly resent any sanitary reforms," declared William Carson, "the Indian masses regard water with aversion and soap with horror."[68] Joséph McCarty extended this derision to poor mestizos, who were, he said, "not over fond of the bath tub."[69]

> *Durango, 1912.* "the mozo [attendant] led us to a large room, with a window opening into a garden, where we could see orange trees and flowers. In the center of the room there was a huge tank, perhaps eight feet square and four feet deep, empty and spotlessly clean, with steps leading down to the bottom. The mozo brought fresh straw mats, two large cotton sheets, rough towels, a little toilet glass with fittings, soap and zacate [fiber], which does service as a sponge. The soap and zacate were in small, tin dishes which float on the water, and are thus near at hand when required. He next pulled out a wooden plug in the side of the tank and a torrent of water gushed in, filling the tank to the height of a man's waist where we could divest ourselves of our clothing. Bob jumped in without ado; but I paused on the top step and dipped in a wary toe to try the water. Finding it only a trifle cooler than body temperature, I too made the plunge and reveled in the soft, greenish clear water, which carries iron and Sulphur. All the cities of Mexico are favored with fine baths, but for delightful water and arrangements I commend 'las Canoas' of Durango."
>
> —Wallace Gillpatrick, 1912
>
> Source: Gillpatrick 1912: 8–9.

Traveler accounts note the existence of baths in Mexico City in the early part of the nineteenth century, but after 1850 artesian wells supplied bathing facilities throughout Mexico.[70] "You will find them everywhere in the large cities," wrote one man in 1886, "and their appointments are first class."[71] Travelers lauded the baths in Veracruz, Orizaba, Xalapa, San Luis Potosí, even the remote northern city of Durango.[72] And while rural dwellers continued to utilize rivers and springs for washing, bathhouses were also increasingly common even in small towns in Mexico. In 1867, for example, James Elton wrote that "the Mexicans are in advance of many European cities with regard to their baths, for in every small town you will find at least one Casa de Baños . . . all of them being clean and neatly kept, and the tariff exceedingly low."[73] Twenty years later Fanny Iglehart noted that "comfortable and luxurious public baths—warm and cold—for all classes exist everywhere."[74] Guadalajara was said to have twenty-six public baths at the turn of the twentieth century. By that time many Mexicans in towns and provincial cities had grown accustomed to bathing frequently in the profusion of public baths, as very few private houses had bathrooms before 1900.

Mexico City, 1886. ". . . while the boys went on to the castle, the girls were left at a corner where a long sign on some low rambling building advertised the Baños de Rosario. They passed through a gateway into a little office with a counter, where tickets were given them in exchange for a moderate sum, and then, following a loosely clad muchacho across the usual garden, they were shown into the bath. It was an immense high place, lighted only from the top, and when the high double door was closed upon them, and Bessie had drawn a huge bolt inside to secure it, they felt somewhat solemn, for all was still within except a sound of rushing water, although in the distance they heard splashing and the laughter of other bathers in other rooms like this one. The floor was brick, the whole space being occupied with a round swimming-place twelve of fifteen feet across. Except a walk two or three feet around it. The edge of the bath was higher on one side than the other, so that running water, coming into the bottom of the bath, while it kept it constantly full, was constantly flowing over the lower margin, where it ran off through a sort of trough. In a moderately dry corner, stood a dressing table with a glass over it, covered with the usual bath implements. There were a couple of chairs by it, with matting in front of them for the feet. In the opposite corner was a shower-bath, and in the space between the clear green reservoir of fresh water, about four feet deep, with a smooth bottom of red brick. It was most inviting. Bessie scorning the steps, had soon plunged in, and was swimming about joyfully in the mild soft water. Helena more cautiously descended the steps and found the water just up to her chin. When they were refreshed rubbed and dressed, they came out again into the office. Bessie thought she was to return the torn-off scraps of yellow tickets which the muchacho had given back to her, but a chorus of assistants exclaimed that these were good for a return ride to the city."

—Edward and Susan Hale, 1893

Source: Hale and Hale 1893: 179.

It was Mexico City, however, that was the capital of bathhouses, and by 1867 "their number was legion."[75] The oldest ones were located in the city center, but new ones sprung up in the late nineteenth century on the outskirts of town, especially along the Paseo de la Reforma, which was planned during the French occupation and often described as the Champs-Élysées of Mexico. Sebastián Pane, the

entrepreneur who pushed forward the groundwater revolution with his drilling rig, was also a leader in the massification of bathing. He opened the famous
"Alberca Pane" ("Pane's Pool") in 1864 on a vacant piece of land on the Paseo de la
Reforma near the statue of Cristobal Colon, and supplied its multiple pools, showers, and baths with three artesian wells. Other similar establishments followed on
its heels, including the Baños Osorio and Baños Blasio right next door, creating
a new genre of bathing establishment—the *balneario*—that had swimming pools
and was as much a waterpark as a bathhouse. Such was the abundance of artesian
groundwater that one of the *balnearios* on the Reforma offered baths for horses,
and the Alberca Pane had a free tank of constantly flowing water outside the building for use by soldiers and the poor.[76] The artesian wells gushed continuously,
nourishing the baths before spilling into the drains, sewers, and canals that led
east to Lake Texcoco. The Albercas Pane and Osorio each had three artesian wells,
and the one that served the main swimming pool of the Pane could fill a water
carrier's jar ninety times a minute.[77] These new luxury bathhouses sometimes had
romantic names, such as "Baños Factor" and "El Harem," and they offered a wide
array of aquatic experiences: "lukewarm baths, hydrotherapeutic baths, russian
baths, and turco-roman baths; the bathhouses had installations suited for practicing swimming."[78]

> *Mexico City, 1886.* "On another Calzada, not far away from the Alameda, were
> the Baños del Recreo, and it was well to take this recreation on their way back to
> the hotel. A friendly woman, mistress of the establishment, sold them tickets at
> a counter in a little room at the entrance of the baths. Passing through this they
> came into a little snug garden, and there was a noise of water rushing, and the
> sounds of merry laughter from the girl's swimming bath. While their baths were
> being prepared they sat in the corridor looking at the flowers, while Tom stole a
> 'ladies delight' for his buttonhole. Then each retired to his or her little cell for a
> refreshing plunge in warm or cold water, after which they were ready for a brisk
> walk home, or to take the street car at the archway."
> —Edward and Susan Hale, 1893
>
> Source: Hale and Hale 1893: 112–13.

As Macías-González (2012) shows us, during the late nineteenth-century administration of President Porfirio Díaz, going to the new bathhouses was an activity
laden with meanings of class, status, and civility. The Alberca Pane was a favorite
social setting of the elite, most notably President Díaz himself, who at the advice
of his physician and friend, Eduardo Liceaga, sought the fortifying and curative
effects of taking the waters. "The Alberca Pane," declared a railway promotional

ALBERCA PANE.
Baños de agua fría, de vapor y ducha.

FIGURE 4. Alberca Pane. Rivera Cambas 1880–1883, vol. 2, p. 284.

travel book in 1894, "is the largest and finest in every respect," with "shower, swim, Roman, Russian and Turkish baths."[79] Much like a European spa, the Alberca Pane offered a wide assortment of cultural and social activities, including gardens, dining rooms, musical performances, swimming lessons, sporting events, hairdressers and barbers, and medical attention. The Alberca Pane's "seductive oriental bath" was especially tailored for the wealthy, with its "beautiful garden and kiosks, carpets, walnut chairs, mirrors, shell-covered furniture" and elaborately tiled pool.[80] Composer, violinist, and popular icon Juventino Rosas gained much of his fame playing Straussian waltzes for the wealthy at the Alberca Pane and the Baños Factor.[81] His song "Junto al Manantial" ["Beside the Spring"] was composed for the birthday party of the wife of the owner of one of the baths.[82] Long weekday afternoons and entire weekend days were spent bathing, eating, socializing, and performing other rituals of class distinction.

As the century progressed, these prominent bathhouses brought together increasingly wider swaths of Mexican society into a hierarchical but still unified space. "There are baths of true luxury," wrote Manuel Rivera Cambas, "and others for social classes with few resources; they are divided in categories aligned with the people that use them, and thus their cost."[83] And the baths were not just for men, although the rules of propriety ensured that men and women occupied entirely

FIGURE 5. *Baños*. De Cuéllar 1889.

separate bathing facilities. In his 1889 novel *Baile y Cochino*, José de Cuéllar tells the story of three sisters from more humble origins who bathed regularly at the Alberca Pane for hygiene, health, and, not least, to mingle with the well-to-do in these new spaces of leisure. Middle-class Mexicans often took the "baths route" (*circuito de baños*) streetcar, operated by the Alberca Pane, paying 50 centavos for a ticket that included entrance to the Russian baths, 25 centavos for the hydrotherapy baths and lukewarm baths, or 12 centavos for the coldwater baths.[84] It was ordinary to see parades of young women with their hair in towels, returning home after their baths on the "Baños" streetcar line.[85]

Poor city dwellers also participated in the public rituals of hygiene and cleanliness. Luxury bathhouses such as the Alberca Pane brought together the middle and upper classes in shared spaces and activities that fortified a notion of belonging to a civilized nation, but they also provided baths for the poor. While the poor could not afford to use the facilities within the bathhouses, they made use of a tank offered by the Alberca Pane free to the public on the street, and in this way experienced the hydraulic opulence and public rituals of hygiene and cleanliness.[86] When the flow of its artesian wells diminished in the 1890s, the Alberca Pane asked for three *mercedes* of additional surface water from the city government to support an expansion of its facilities and "a reduced price to poor people," which was awarded on the basis that the bathhouse provided a "benefit to public hygiene and health."[87]

In the cities of Europe and North America, municipal governments built bathhouses beginning in the last decades of the twentieth century with the purpose of promoting cleanliness among the working class,[88] and the government of Mexico City made similar plans, beginning in 1881, to provide public bathhouses in each of the neediest zones of the four cardinal directions in the city, where, city *regidor* (alderman) Ignacio Toro declared, "the poor bathe and wash their clothes in the canals and ditches, which is manifestly unhygienic" and encourages "immoral" public nudity.[89] The public works commission explained to the *regidores* that public bathhouses had "been built with brilliant success in England; and in France a fund of 600,000 francs had been created for the same purpose."[90]

"Thanks to notions of popular hygiene that spread more and more every day; thanks to the progress and aspiration of our civilization; we vehemently support, in order to secure greater well-being, the widespread use of baths in our city. You can see a great number of people of the lowest classes that make use of the canals and even drainage ditches to wash their clothes and bodies, mostly on their days off. If they do it in those places that are full of germs of diseases that spread through the waters, it is because they have nowhere else to do so for free."
—Mexico City Public Works Commission, 1895

Source: AHCM, Policía en General, Vol. 3639, Exp. 1064 (1895).

These planned municipal bathhouses were to provide immersion baths as well as laundry facilities, and in 1887 eleven thousand pesos were authorized for the construction of the first such facility in Mexico, dedicated to serving women. The money never appeared, however, and the project ran afoul due to a number of other problems. The bathhouses required a quarter of a city block, and such properties were unavailable, as was the water needed for the bathhouses. In 1891 a new proposal was submitted to the city council by Antonio Torres for a smaller, less costly bathhouse, with multiple entrances, fewer laundry basins, and more bathtubs. Torres was particularly worried about offering too many sinks, because "it would enable the paid washerwomen to go there rather than those truly in need."[91] By June of 1891 the city had agreed to purchase a lot in the Colonia Morelos for the bathhouse, and contracted Guillermo Paterson to drill an artesian well to supply the water.

In the end, however, the bathhouse was not built. Paterson assured the city that water had been found in a nearby lot at 100 meters, but only found the liquid at 222 meters after drilling for seven months, more than doubling the costs. The process of purchasing the lot for the bathhouse dragged on and on, and the money for the construction was never delivered. The burbling artesian well became a fountain for the neighborhood, and someone installed themselves as caretaker of the well and the property.[92]

"In April to June, when the heat is greatest in the capital, you see the masses of inhabitants going to the pools with the most extraordinary dedication. But not only at that time: they go most of the year, because in addition to the pools fed by artesian wells there are also showers, russian [sic] baths and all the others that are used for medicine or recreation. In June more than 40,000 bathers go to the pools, arriving on the trams, in cars, on foot or horseback, happy caravans going to those wonderful leafy spaces where the trees, the landscape and the company call their attention. There are beautiful young women with their hair down and adorned with flowers, the throngs of vendors selling snacks, and often enthusiastic musicians. . ."

—Manuel Rivera Cambas, c. 1880

Source: Rivera Cambas 1880–1883, vol. 2: 285.

The Feast Day of Saint John the Baptist, June 24, was a crucial event in the conformation of water culture in modern Mexico City. On this warm summer day Mexicans traditionally celebrated the Catholic association between purification and water by visiting a bathhouse or a nearby river or spring. A common saying was that bathing on that day would give "beauty to the maiden, vigor to the matron, and freshness to the old maid."[93] This custom grew more elaborate with the hydraulic opulence of artesian wells and the proliferation of bathhouses such as the Alberca Pane in the late nineteenth century. The baths and pools filled to capacity on that day with children and adults enjoying the water much more in pursuit of ludic than spiritual ends.

"From the first ring of the church bells in the morning [of Día de San Juan], those who were heading to the baths took to the streets happily singing. . . . Some took the road to Chapultepec in wagons and trams, others headed off to the different baths around the city, which were swept, washed and decorated with willow branches around the patios, doors and windows, and sparkling everywhere with decorations. . . . The energy in the baths were extraordinary, and the general happiness was increased by the sounds of the musicians. The bathers reflected this emotion with their shouts and laughter, the splashes made every time one of them dove in. . . . It was a custom in all the baths to give away fruit, soap and sponges. . ."

—Antonio García Cubas, 1904

Source: García Cubas 1904: 374.

The *día de San Juan,* like other festival days, was intensely social. Bathhouse owners adorned their buildings with plants, banners, flags, and other decorations, and vendors set up carts and stands catering to the crowds that descended on the baths. Grooming items were available everywhere, and the bathhouses themselves offered gifts of soap and small scrubbing pads made of cactus fiber (*estropajos,* made of *ixtle*) to their clients. In addition, all sorts of food was available on the streets outside the bathhouses and pools.

> "In one tank one hundred and fifty or more bathers may be seen at once, throwing themselves head first, diving and swimming, or standing half submerged, or perhaps jumping from the spring-board. To all these gyrations add the screams of the multitude, the shrieks of the bathers and the people on shoe selling a thousand and one articles beneath the rays of a scorching sun, to complete the scene. Though many pursuits and avocations are carried on, the dominating and supreme desire of the crowd is to get wet."
>
> —Fanny Iglehart, 1887
>
> Source: Iglehart 1887: 275.

Bathing in the bathhouses of late nineteenth-century Mexico was an activity in which a wide swath of society participated, and it helped generate a sense of the nation rooted in traditional customs such as those of the *día de San Juan* as well as new practices of civilized cosmopolitanism. Artesian wells allowed sumptuous practices of immersion, once a symbol of European culture and a privilege of the elite, to soak far down into Mexican society. Bathing establishments remained segregated along class lines, but the availability of bathhouses at all price points made it possible to imagine that all Mexicans were unified as a nation that bathed together. This vision placed Mexico on the same level as Europe, and was bolstered by an evolutionary narrative that seized upon bathing and cleanliness as signs of civilization. As one intellectual from the time stated, "baths with the luxury and dimensions that the ones on Bucareli and Reforma already have in Mexico, are an undeniable proof of advanced civilization."[94]

While in the 1840s Ignacio Cumplido described the rustic baths of Guadalajara as desirable for their simplicity, and derided the Roman baths as decadent, by the 1880s Rivera Cambas cast this classical opulence in a positive light, stressing the continuities with modern Mexico. "Although our civilization has not refined its sense of taste to the degree that it was in the era of Cicero, we nevertheless have beautiful medicinal and recreational baths," he wrote.[95] "Someone once said, and was right," declared Prantl and Groso, "that the level of culture of a people is manifested in the number and quality of its bathhouses." Mexico was still at a middling level in these terms, they continued, but nonetheless "fulfilled the requirements of cleanliness, comfort and hygiene demanded by such a civilized metropolis as ours."[96]

Baths and bathing in Porfirian Mexico evoked classical civilizations by borrowing Roman, Greek, Moorish, and Aztec elements. As Mexico's economy and state grew in the last decades of the nineteenth century, the emergent Mexican bourgeoisie pushed back against unilineal evolutionary thinking that cast the country as barbaric and "backward," by claiming its own classical tradition. Just as European countries took the mantle of civilization from Greece and Rome, Mexican intellectuals such as Antonio Peñafiel made an antiquarian effort to recover Mexico's roots in the complex societies of Mesoamerica.[97] National histories from this time drew continuities to the grandeur of Tenochtitlán as well as Spain,[98] a narrative that was projected internationally in settings such as the Universal Expositions in Paris in 1889 and 1900, for which Peñafiel designed a "Mexican Pavilion" in neo-Aztec architectural style.[99] At home in Mexico City, statues of Cuauhtémoc, the last Aztec emperor, and two "Indios Verdes" graced the Paseo de la Reforma along with those of Christopher Columbus and King Carlos IV of Spain.

"Who says our beloved Mexico is not civilized? What a crass mistake! And is there anyone in Mexico who complains of an incurable disease? No one, no one, no one. Who is going to get sick in Paradise? Mexico is the Garden of Eden with those baths."

—Artemio de Valle Arizpe, 1946

Source: De Valle Arizpe 1946: 156.

Mexico's bathhouses evoked the splendor, opulence, and refinement of the classical Mediterranean world. The "Coliseo Nuevo" (New Coliseum), founded around 1850 by an expatriate Italian general, was renamed "The Harem" soon after.[100] In 1887 the Alberca Pane installed a "Turkish-Roman" bath they called *El Hammam* (the Arabic name for the classical Islamic bathhouse) that offered a sumptuous Roman sequence of water encounters: the tepidarium, the caldarium, the laconicum, the alipterium, the lavatorium, and finally "showers of different temperatures" or a cold plunge bath.[101] The "Turkish bath," introduced around 1900 to Mexico City, offered a hot steam treatment quite similar to that of the *temazcal* before it, but for an elite clientele and with very different connotations.[102]

Mexico's civilized trajectory could also be clearly seen in the Bosque de Chapultepec. The Bosque was widely known to be a hunting ground and park for the Aztec rulers, and the famed springs, the pools that collected their water, and the aqueduct that delivered it to the city center were all associated with the Aztecs, who first built the water system. The smallest spring-fed pool was the source of

FIGURE 6. *Baños* de Chapultepec. Michaud 1874. With permission of Universidad Nacional Autónoma de México, Instituto de Investigaciones Estéticas, Archivo Fotográfico Manuel Toussaint, Colección Julio Michaud.

water for the aqueduct, and was the oldest. Travelers and locals viewed it as "an interesting relic of Moctezuma's glory," referring to it as "Moctezuma's Bath" and "Moctezuma's Pool," and believing that it "was probably used by him."[103] In fact, the pool was built as a reservoir for the water that flowed to the aqueduct, and it is highly unlikely that it was used by Moctezuma for bathing or swimming given that it was the water source for the city, and that people "bathed" in *temazcales*. The idea that this was an Aztec bath may have been influenced by a famous pool in the hills near Texcoco, known as Nezahualcóyotl's Baths, a common stop on the itineraries of travelers to Mexico since the colonial period that gained renewed fame with the growth of tourism in the late nineteenth century.

By the 1860s the owners of Chapultepec's Alberca Grande (also known as the "Alberca de los Nadadores" or the "Swimming Pool") had built a bathhouse to serve the public.[104] This was the most popular swimming pool for city dwellers of some means until the Alberca Pane and its neighbors opened up on Reforma. The bathhouse was built in a neoclassical architecture, and decorated "in the style of Pompeii," with a large swimming pool fed by the springs and smaller, private pools and rooms that received the water from the Alberca Grande.[105] There were gardens with sandy walkways shaded by enormous *ahuehuete* (cypress) trees. Antonio García Cubas describes (with his typical thesaurical largesse) a "rich and endless spring almost overflowing the pool that bounded its transparent waters, where the good swimmers showed off their prowess, jumping off the high guardrails into the liquid to catch a silver coin as it sank, or to lie beneath the tree roots in the water to display their ability to hold their breath as well as the best divers."[106] One traveler called a swimmer "a swarthy son of Aztecs," reinforcing the popular narrative about the classical Mesoamerican origins to bathing in Mexico.[107]

The Alberca de los Nadadores operated from the 1860s until about 1880, when the profusion of artesian wells around the springs reduced their water levels so much that the city stepped in to purchase the title to all the springwaters and channel them to the aqueduct that led downtown. So much water was extracted from the subsoil in Chapultepec that the ancient *ahuehuete* trees—also associated with the Aztecs—began to die, prompting caretakers to ask the city for a concession of springwater to irrigate them.[108] By the time García Cubas wrote his memoirs in 1904, the pool was dry and already eulogized as the remnant of a noble and hygienic indigenous civilization. "Montezuma's bath still stands," wrote Crawford in 1899, "a charming bit of ruins."[109]

At first only the relatively wealthy could afford the sumptuous, novel encounters with water offered at the city's new bathhouses. But as we will see in the next chapter, as water became more available greater numbers of bathhouses opened, and poorer people had more access to swimming pools, *placeres,* and showers. Like many other elite practices and symbols, the new forms of bathing were slowly adopted by the masses. By the 1920s Mexico's bathhouses served a mostly popular clientele, as the opulence of water, confirmed by the Xochimilco aqueduct, trickled down through society. Partially as a result of the massification of social bathing, the correct way to wash the body was recast by sanitarians from the public bath to the private domestic shower, which was in turn promoted by the federal government.

CONCLUSIONS

In 1850 bathhouses and open-air bathing sites were well attended throughout Mexico, and swimming was a popular pastime for many Mexicans. After that, however, new sources of groundwater facilitated a grand expansion of these activities. In Mexico City, swimming pools and bathhouses opened in new neighborhoods along the Paseo de la Reforma, where bathing took on a modern, cosmopolitan air. Chapultepec Park itself had a swimming pool and bathhouse. Social bathing for fun and fitness grew in popularity and many new businesses were opened that offered new watery experiences: saunas and steambaths, hot springs and swimming pools.

This was a significant shift in the encounter that people had with waters: from steaming and washing to soaking and showering. There was a cultural resignification of cleanliness, which was increasingly defined in terms of hygiene and linked to concepts of civilization and progress. The *temazcal* was cast aside as a tradition practiced by poor and indigenous people, and new bathing practices

FIGURE 7. El Pozo "Pimentel," La "Colonia de la Condesa." *El Mundo Ilustrado* 2, no. 12 (1906). With permission of the Hemeroteca Nacional de México, Fondo Reservado.

of immersion—and even more so, showering—came to be seen as modern and desirable. Sidney Mintz calls this "extensification," a cultural process in which the poor emulate the practices of the wealthy.[110] But just when the humble residents of the capital gained access to bathhouses, the rich built private bathrooms in their homes, which became commonplace in the bourgeois *colonias* such as the Roma and the Condesa that expanded on the western side of the city. This shift was enabled by groundwater: the Condesa neighborhood was supplied only by artesian wells when building began around 1905.[111]

During the twentieth century, the expansion of urban water systems increasingly provided water to household bathrooms as well as collective bathhouses. Newly built hydraulic infrastructure was considered evidence of both Mexico's status as a civilized nation and the power and authority of the Mexican state.[112] After the Xochimilco aqueduct was completed around 1910, houses in the new neighborhoods in the wealthy areas of Mexico City were connected to the city water grid and plumbed for showers, a trend that would continue through the twentieth century. Groundwater is of course not limitless, and the boom of artesian wells was relatively brief—1850 to 1900, more or less. Nevertheless, the groundwater boom produced habits of water use that, even after the aquifers were depleted and the artesian wells trickled out, lived on to motivate the ceaseless twentieth-century drive to build ever-more-encompassing works to supply universal, frequent, individualized household baths with uniform, public water.

6

Chemistry, Biology, and the
Heterogeneity of Modern Waters

Scholars of water argue that large-scale public water systems built with new engineering techniques in the late nineteenth and early twentieth centuries created "new water"—uniform, homogeneous, and public.[1] What is not often remembered, however, is that the creation of new water depended on a continuing appraisal of the heterogeneity and specificity of waters. In building public works, engineers had to confront the specific details of particular water sources, such as location, origin, flow rate, mineral content, and other variables. Chemists and biologists working to ensure that public water met uniform health standards needed to identify and measure the biological and mineral contents of these waters. Although the groundwater that supplied the bathhouses of Mexico was not hot, not highly mineralized, and did not spring to the surface by itself, it was nevertheless incorporated into existing classifications of those kinds of waters, in recognition of its specificity and its relation to other waters. Scientists analyzed the mineral contents of artesian wellwaters, constructed theories about their geological origins, looked for microbes, and reached the conclusion that they were perfectly suitable for inclusion in homogeneous public water.

Ideas about the heterogeneity of waters also evolved due to important developments within science. At first, those who studied water were mostly chemists, but after about 1860, biologists equipped with more powerful microscopes identified organisms that caused diseases that were previously thought to derive from the waters themselves, or from the gases that emanated from them.[2] With the rise of bacteriology, it became evident that cholera, yellow fever, malaria, and other diseases did not result from physical aspects of the climate, environment, and geology, but rather from organisms that grew in water. Hygiene and sanitation in public health squared off against these bacteria, in an effort to sterilize and sanitize public water. In this process of creating and imposing uniform standards for public water, the virtues of heterogeneous waters were sometimes forgotten.

But they were not forgotten for long. Despite the expansion of infrastructure and the shift to biological understandings of health and disease, the idea of a homogeneous, "public" water never completely dominated, neither in popular nor scientific minds. Mexico's medical community was filled with pharmacists and chemists who continued to research the content and therapeutic qualities of Mexico's many waters, and in particular, its groundwater and mineral springs. Leopoldo Río de la Loza was a central figure in the resilience of physical-chemical approaches to water and health, who, from his chair in the National Academy of Medicine and National School of Medicine, directed research and trained generations of scholars. One of his students, Eduardo Liceaga, pioneered bacteriological approaches to health in Mexico, introducing vaccinations and addressing outbreaks of yellow fever through quarantine. Liceaga rose in prominence to direct the National Health Council and other medical institutions during the ascent of bacteriology, but he maintained a deep interest in the therapeutic uses of waters, especially the physiological effects of baths and showers, and he promoted research on bathing at the National School of Medicine, at the National General Hospital that he designed and built, and at the mineral hot springs of Peñón de los Baños.

The same economic growth that spurred the construction of infrastructure and bathhouses in the Porfiriato also promoted the development of hot springs and mineral springs into medical facilities and business. Even though hot springs bathhouses such as Peñón fell into decay during the early Republican period, most Mexicans continued to hold deep-seated beliefs about the medical benefits of mineral water bathing. The idea that bathing in and drinking mineral waters was medicinal and therapeutic enjoyed a resurgence with the popular "hydropathy" movement in the 1840s and again, in a more elite scientific form, in the late nineteenth and early twentieth centuries, even with the rising hegemony of bacteriology. A new bathhouse was built at Peñón, and Liceaga himself opened a bathhouse in Villa de Guadalupe. Furthermore, as the nineteenth century progressed, relaxation and recreation were added to the benefits ascribed to water therapy.[3] So while uniform public water consolidated its presence in Mexico after 1850, it was accompanied by a booming science and business of heterogeneous waters.

THE SCIENCE OF GROUNDWATER: CHEMISTRY AND BIOLOGY IN BALANCE

By the 1850s the rush was on in Mexico City to drill for water. The boom of artesian wells raised geological questions about groundwater. Where was it located? How did it flow? Was it connected to surface water? Sebastián Pane and his partner D. Augustin Molteni provided material from one of the first well bores to Leopoldo Río de la Loza, a chemist who studied waters, so that he could sketch the strata underlying the Valley of Mexico.[4] The Ildefonso brothers and Ignacio Ortiz de

Zarate did the same when they opened a well at the Casa de la Moneda, just off the Zocalo in 1871.[5] Technological improvements allowed engineers to discover ever-deeper water-bearing strata—at 52 meters in 1858, 105 meters in 1863, and up to 234 meters below the surface in the case of the well sunk by Carlos Pérez Rívas near the Military Hospital. Wells sunk far from the city center encountered the same strata as those outlined by the geological studies, but at different depths.[6] These wells usually tapped the third aquifer from the surface, which according to Río de la Loza held the best quality water, but deeper and shallower ones were also commonly used.

Because of artesian wells, the water supply almost doubled by 1858; by 1883 it almost tripled.[7] At the same time, however, aquifer water was an unknown substance, and there was no information about where it came from, how much there was, its mineral content and quality, its relation to surface waters, or the effects of extracting it from the ground. No one knew if it was safe to drink. Mexico City's varied waters had always been conceived of as unique, their qualities associated with the places they emerged. The springs of Santa Fe were "softer," "lighter," and "thinner" than the springs at Chapultepec; the springs in the Desierto de los Leones were found to be even purer, more "crystalline." Well-drilling in the 1850s introduced new waters into the lives of Mexicans, but where did the artesian waters come from, and how did they compare to the known waters?

Noel Coley and others have shown that the modern disciplines of chemistry and medicine were formed to a significant degree through the analysis and replication of mineral waters, and this can be seen in Mexico as well.[8] How scientists approached the question of health and water changed dramatically between when the first artesian well was drilled around 1850 and the completion of the Xochimilco springs aqueduct in 1910, due to a conceptual paradigm shift ushered in by the identification of microorganisms and their linkage to fermentation, putrefaction, foul smells, and disease. For millennia health had been seen as an organism's relation to the qualities and elements in its environment ("airs, waters, places," in the climatology established by Hippocrates).[9] Waters were animate; they had agencies that were described as "virtues" inherent to them. The science of chemistry reshaped this idea by isolating the efficacious chemical elements in the water that generated pathologies and therapies, and recasting the water itself as an inert medium. When, around 1880, the understanding of health moved toward the presence or absence of harmful microscopic organisms in the environment, the material agency of waters was reassigned to the organisms, further robbing the waters themselves of agency. Despite this, the view of water as an inanimate, uniform medium for biological agents never took complete hold. In fact, water culture in Mexico was remarkably conservative, retaining the ancient focus on the relation between bodies and local environments. Even while microbiology changed understandings of health and hygiene, doctors and laypeople continued to view mineral waters as important agents of well-being.

The shifting coexistence of medical scientific paradigms is exemplified in the work of two of Mexico's most important scientists in the nineteenth century: Leopoldo Río de la Loza (1807–1876) and Eduardo Liceaga (1839–1920). Leopoldo Río de la Loza was a chemist and professor at the National School of Medicine. Born into a family of chemists, his studies of chemistry and medicine in the university launched him on a successful career as an academic. At the same time, he founded a number of chemical factories and came to own three pharmacies, or *boticas*: "La Portacoeli," "La Botica de Vanegas," and "La Merced." *Boticas* often sold mineral waters and the salts that were derived from them as treatments, and some chemists and pharmacists replicated those in their laboratories. L. Pauer, for example, produced copies of mineral waters from Vichy, Spa, Carlsbad, and other famed European watering places in the Botica del Refugio, on Espiritu Santo street in downtown Mexico City.[10]

Río de la Loza was part of this process of the constitution of science far from Europe. He was instrumental in compiling the *Farmacopea Mexicana* (1846) and the *Nueva Farmacopea Mexicana* (1874), both with long sections of recipes for mineral and medicinal waters. He conducted numerous studies of waters during the mid- and late nineteenth century that were aimed at identifying the hygienic and therapeutic effects of waters in Mexico. In 1840 he published a study of the effects of lead pipes on water quality in Mexico City; in 1844 he was called upon to do an analysis of the waters of Peñón de los Baños and later published a study of the mineral waters of Atotonilco.[11] In 1847 he was commissioned to study the Xacopinca spring; in 1858 and 1863 he published analyses of artesian wells. In 1869 he served on the Comisión Sobre las Aguas Potables de Mexico, and later published a study of springs and potable water in Teotihuacan.[12] He was a member of the Sanitation Board (Junta de Sanidad) of Mexico City, and later of the National Health Council (Consejo Superior de Salubridad), the independence-period heir to the colonial police of public health.[13]

Río de la Loza led the geological and chemical study of groundwater in mid-nineteenth-century Mexico. He quickly developed a close relationship with Sebastián Pane, describing his drill as the "exploratory probe" for his studies. In October 1858, at the beginning of the boom in artesian wells, Pane sank a well at #2 Calle de Santa Catarina, northwest of the Zocalo. Río de la Loza and fellow chemist Ernesto Craveri studied the soils extracted by the drill, compared them to samples from other wells drilled at that time, and generated an image of the geological formations underlying Mexico City. "Because of this information," they wrote, "we believe that in this valley, at the depth of fifty meters more or less, there are three strata of water that have the conditions necessary for supplying artesian wells."[14] They concluding that each of these aquifers held its own kind of water, hydrostatic pressure of these aquifers varied, and depending on their depth the artesian wells produced between 720 and 2,880 barrels of water each day.[15] Río de la Loza's geological studies helped promote the assumption of opulence of groundwater, for while many at the time assumed that the water from the new artesian wells would run out, he declared with scientific certainty that the artesian wells were "permanent."[16]

Another pressing question about artesian wells concerned public health and the chemical composition of the waters. Would they harm? Could they heal? When the first artesian wells in Mexico City bubbled forth, these strange new waters were not well received by the wealthy households they served. People claimed that the artesian water upset their stomachs and made their hair fall out. Some of the first wells produced salty water because the engineers did not prevent surface water from mixing with that drawn from deeper aquifers. City dwellers complained that water from some of the wells, such as those on the Calle de Los Cordobanes (today, Calle de las Donceles) and the Aduana (today, Calle 5 de Febrero), was "azufrosa" (sulfurous) or "hedionda" (stinky), because of a sulfurous smell that reminded them of hot springs.[17]

Río de la Loza and Craveri were commissioned by Pane and the Chamber of Industry of Mexico City to determine the healthfulness of artesian waters and compare them to others in the Valley of Mexico. Like the mass of people without scientific training—often referred to by scientists as the "*vulgo*"—scientists in 1850 began with the assumption of miasmatic theory that foul smells in air or water were bad for health, and understood these miasmas in terms of chemistry and climate, not microbes or bacteria.[18] The artesian wellwater smelled badly, and Río de la Loza sought to identify the minerals that caused the smell and to understand their effects on the "economy" of the body.

Río de la Loza concluded that the artesian waters were better for the health of the public than others in the Valley. The sulfurous smell that people noted was actually a harmless gaseous hydrocarbon that would lessen over the life of the well, and would evaporate from the water if left standing. Popular ideas that groundwater caused people's hair to fall out were simply unscientific and wrong. "When some inhabitants of Mexico City," he wrote in 1863, "who are used to drinking the so-called 'thin' water [agua delgada], change it for the 'thick' [agua gorda], their digestion will suffer for a few days, more or less."[19] His analysis showed that the "thick water" had more dissolved minerals than groundwater, and argued that it was the calcium and magnesium, as well as the salts, that caused these digestive problems. But, he argued, the artesian wells produced clean water with relatively little dissolved minerals. Artesian water, Río de la Loza insisted, was not bad for people, and to protect public health hygienists should instead take aim at social and cultural factors such as "habits, changes in location and dwelling, etc."[20] Río de la Loza recommended using water from the third aquifer from the surface, which was of better quality than the first water-bearing strata that was cheaper to access.[21]

Río de la Loza was at the forefront of medicine and public health in Mexico in the mid-nineteenth century. He lived to see John Snow's discovery in 1854 that cholera and other diseases were transmitted through London's groundwater, and that it was germs rather than miasmatic gases and airs that caused disease. But he died in 1876, just two years before Pasteur published his landmark study *Les Microbes Organisés,* which sparked a hot debate at the 1878 Hygiene Congress in Mexico City between the established medical tradition and the new adherents to microbiology. Río de la Loza left the National School of Medicine solidly oriented toward chemical analysis, but in the following decades medicine and public health would slowly incorporate biology.

This transition can be seen in the life and work of Río de la Loza's most notable student, Eduardo Liceaga. Liceaga passed his medical exam in 1866, and went on to be a leader in science, health, and medicine in Mexico until the twentieth century. In 1887 and 1888 he toured the capitals of Europe, visiting hospitals and the Pasteur Institute in France, and returned to Mexico with materials for vaccinations against rabies and, having visited sewer and potable water systems, a keen interest in public works, hygiene, and water. He adopted the bacteriological approach and created modern institutions that characterized health in the twentieth century, serving twice as president of the National Academy of Medicine. He was an important political figure who also held the presidency of the National Health Council, helped write the 1891 Sanitary Code, oversaw the construction of the National General Hospital (1905), led the prophylactic effort to identify and quarantine yellow fever in Mexico's port cities, and founded Mexico's National Bacteriological Institute in 1905. As the personal doctor of President Porfirio Díaz, who ruled during most of the period between 1876 and 1910, his access to power was guaranteed.

Despite Liceaga's remarkable success in promoting microbiological approaches to health, there was no moment in the history of medicine in Mexico that marked an abrupt break from earlier approaches to health that focused on climate and environment. As Eric Jennings (2006) has shown in his study of hot springs in France and its colonies, the turn away from climatology was a slow process of incremental change as centuries-old views of health bent, adapted, but only sometimes broke under the force of the new paradigm of microbiology. According to Paul Ross (2009), doctors in Porfirian Mexico continued to "explain disease as a complex relationship between local environmental conditions (especially miasmas) and individual predisposition," rather than a result of tiny organisms.[22] So while Liceaga ushered in the bacteriological approach to health in Mexico, he was also a leading proponent of therapeutic bathing and mineral waters.

Water continued to be a principal concern of doctors and health officials in Mexico City. The proliferation of artesian wells in the 1850s and 1860s focused attention on the quality of groundwater at a time when scientists were still mostly focused on chemical virtues. The huge increase in water supply generated by these artesian wells only accentuated the problem of stagnant and noxious waters, which formed wetlands around uncapped wells and sluggish pools in the city's drainage canals. By the 1880s the stench of nearby Lake Texcoco, which received the city's effluent, was unbearable to many among the educated elite who, informed by discussions of hygiene, sought to create a more sanitary city.

In 1882, the National Academy of Medicine commissioned a study of "the influence of waters for domestic use on the public health of the Capital." The result, Antonio Peñafiel's *Memoria de las Aguas Potables de la Capital de Mexico*, shows the evolving balance between chemistry and biology, and climate and microorganisms, in ideas about waters, health, and cleanliness. In that document, chemical analysis was still paramount, but following the emergent emphasis on microbiology the object of study had turned to the putrefaction of organic material in the water, caused by the explosive growth of microscopic plants and animals "in the millions."[23] Air, water, and organic material were the key ingredients for this fermentation, a process which consumes oxygen and produces carbonic acid and ammonia. Peñafiel followed the work of Pasteur, but his analysis of water and health pointed him back to the chemistry of waters—to the presence of carbonic acid and ammonia as identifiable markers of infection in the waters of the Valley of Mexico. "Pasteur has not finished building his theory, but we can seize on the most prominent and visible results of these vital, chemical actions," Peñafiel suggested.[24] He offered a discussion of microscopic analysis of bacteria in water, but in practice gauged the relative chemical purity of waters in the Valley of Mexico by the presence or absence of macroscopic living organisms such as fish and snails. The microbiology of contagion was still something of a black box in Peñafiel's climatological method and theory.

While climatological perspectives may have held their own in discussions of potable water between 1880 and 1920, they actually grew in prominence overall due to their role in the dramatic growth of the business of bathing. Water was neither just the medium through which microbiological threats to public health came into contact with people, nor the substance that could be used to wash those threats away. Waters themselves were increasingly considered crucial for both hygiene and therapy among doctors and the public, despite the emergence of homogeneous public water. This resurgence of the conceptual specificity and multiplicity of waters, and of the notion that waters were agents in a climatologically informed health system, unfolded in the practices and places of the bath.

BATHING FOR HEALTH: THERAPY AND HYGIENE

"There is nothing like water; it will cure all complaints but poverty, and heal all wounds but sorrow! Do you find yourself afflicted in mind, melancholy, or disposed not to hear mass? Drink water, and bathe yourself in the river. Are you stung by a scorpion? Bathe the wound in water: and for the bite of a rattlesnake it is equally efficacious. I am sixty-nine years of age, and for 35 of these I have been a water carrier; and during the whole of that time I have preserved my health by drinking water! There is nothing like water for the head or toothache. Warm water however swells the stomach; but cold water, that is the thing—used three times, it is a remedy for soul and body: for coughs, colds, rheums, colic, and in short every other complaint whatsoever, a liquor for angels to drink with pleasure and advantage."
—Water carrier, Lagos de Moreno, Jalisco, 1829

Source: Hardy 1829: 497.

During the Porfiriato, artesian groundwater and the idea of hydraulic opulence encouraged the massification of bathing for cleanliness as well as social and ludic ends, but many people continued to treat maladies with water. As we have seen, the quality of the particular water was often seen to be the curing agent, and water cures employed the entire range of waters, from pure, fresh springwaters to the most heavily mineralized hot springs, as well as seawater.[25] Each of these waters was thought to have particular properties that made it useful for treating certain diseases. Hot springs were especially important, and the study of hot springs in the late eighteenth century and nineteenth centuries was focused on generating classifications for the mineral contents of those springs and their utility in treating different conditions. The physical action of water on the body was also thought to have therapeutic effects, and a plethora of showers, baths, and drinking schedules were designed to apply water to different parts of the body. For these applications, the content of the water was not as important, and the showers, baths, and other applications utilized whatever water source was at hand.

Therapeutic bathing was practiced in different forms by ordinary folks across Mexico. In the 1820s medical doctor Robert Hardy toured northern and western Mexico, and reported with ethnographic detail on regionally specific popular customs and ideas about water. Hot springs were considered by people in northern Mexico to be curative, but bathing in cold waters was not.[26] In Sonora, snakebites were washed with cold water, while immersion was seen to be harmful for people with colds. Those with smallpox and measles stopped washing altogether for forty days. When Hardy ordered a bath for a sick young girl, her father swore that "he had not closed his eyes during the whole night, as he thought it was not possible that his daughter should survive the washing."[27] The doctor remarked that in this region there was "a kind of superstitious awe felt by the natives in regard to ablutions."[28] On the other hand, in Lagos de Moreno, Jalisco, water was promoted as something of a cure-all, at least by a man selling the water. This same strong idea of the beneficial virtues of water was observed fifteen years later in Lagos by Albert Gilliam, who wrote that the patient "was directed first to bathe seven times, and that afterwards [the doctor] gave him some roots, of which he made teas to drink."[29] Like Hardy, Gilliam considered this to be "superstition" rather than science.

Popular water cures were influenced by transnational trends in medicinal water culture, as was the idea held by others that these cures were unscientific. The "hydropathy" of Vincent Priessnitz was especially influential. Priessnitz was an Austrian with no medical training who established a treatment center on his farm in 1826. He practiced a regime of coldwater showers and wraps, together with a diet of simple regional food, which by 1840 had become famous enough to attract visitors from all ranks of society, and from as far away as England and the United States.[30] Soon hydropathic treatment centers were popping up around Europe and the Americas. Alistair Durie shows how these "hydros," with their abstemious and ascetic qualities, gained adherents among a middle class that turned away from the perceived decadence of spas.[31]

Hydropathy also gathered followers in Mexico during the 1840s and 1850s. But in addition to the idea that it was a more respectable water cure than the spa, Mexican proponents of hydropathy argued it was more popular and democratic. The hydropathic regimen utilized cold, pure water flowing directly from the source for bathing and drinking.[32] According to this perspective, pharmaceuticals were damaging; no other substance than cold pure water was medicinal, not even mineral water.[33] In Europe at that time, public access to hot springs was increasingly restricted by the doctors who made their business with them. Hydropathy encouraged people to make their own cure, as cold water, unlike hot or mineral water, was universally available. In the 1840s hydropathy enjoyed a wave of popularity in Guanajuato, Guadalajara, Silao, Morelia, and Mexico City, promoted by Emeterio Sáez de Heredia and José Nogueras, both priests from overseas.[34] It became so popular in Guadalajara that 150 citizens petitioned the city government to formally endorse the treatment.[35]

Hydropathy was fiercely debated and hydropaths were pitted against the scientific medical industry. Emeterio Sáez explicitly avoided scientific language in an effort to "be understood by the poor and ignorant."[36] Instead, he presented his water cure in religious terms reminiscent of studies of mineral springs from the eighteenth century in Spain and Mexico.[37] While popular notions of health were still rooted in this language, by the mid-nineteenth century it had been purged from scientific discourse, a turn strengthened further by growing Liberal anticlericalism during the late 1840s and 1850s. Sáez attacked scientific medicine, fulminating against the "ambition of glory and fortune" that motivated doctors. He argued that the medical profession attacked hydropathy because it "robbed them of their science . . . the pharmacist trembles, fearing for his business and his drugs."[38] José Nogueras, also a priest, echoed this therapeutic populism when he told his readers "do not expect the flowery language of the classroom, nor the elegant style common in prologues: I will speak to everybody, following the path of nature."[39]

Medical doctors rejected hydropathy as a "vulgar," empirical approach that lacked scientific theory of disease and knowledge of anatomy, and they labeled its proponents "charlatans."[40] At the behest of doctor Juan Manuel González Urueña, José Nogueras was ordered by the government of the City of Mexico to stop practicing hydropathy, although Nogueras later got the federal government to lift the ban. While most doctors in Mexico looked upon those who practiced hydropathy to be quacks, they were at the same time careful to recognize "hydrotherapy," or the scientific use of water in medicine, as legitimate.[41] At stake was the conceptual and practical system by which water was applied to medical ends, not the status of water as a useful substance for medicine. As a result of the hydropathy episode, however, Mexican doctors turned away from water treatments until the 1870s, when the Military Hospital, the Hospital de San Lucas, and some of the

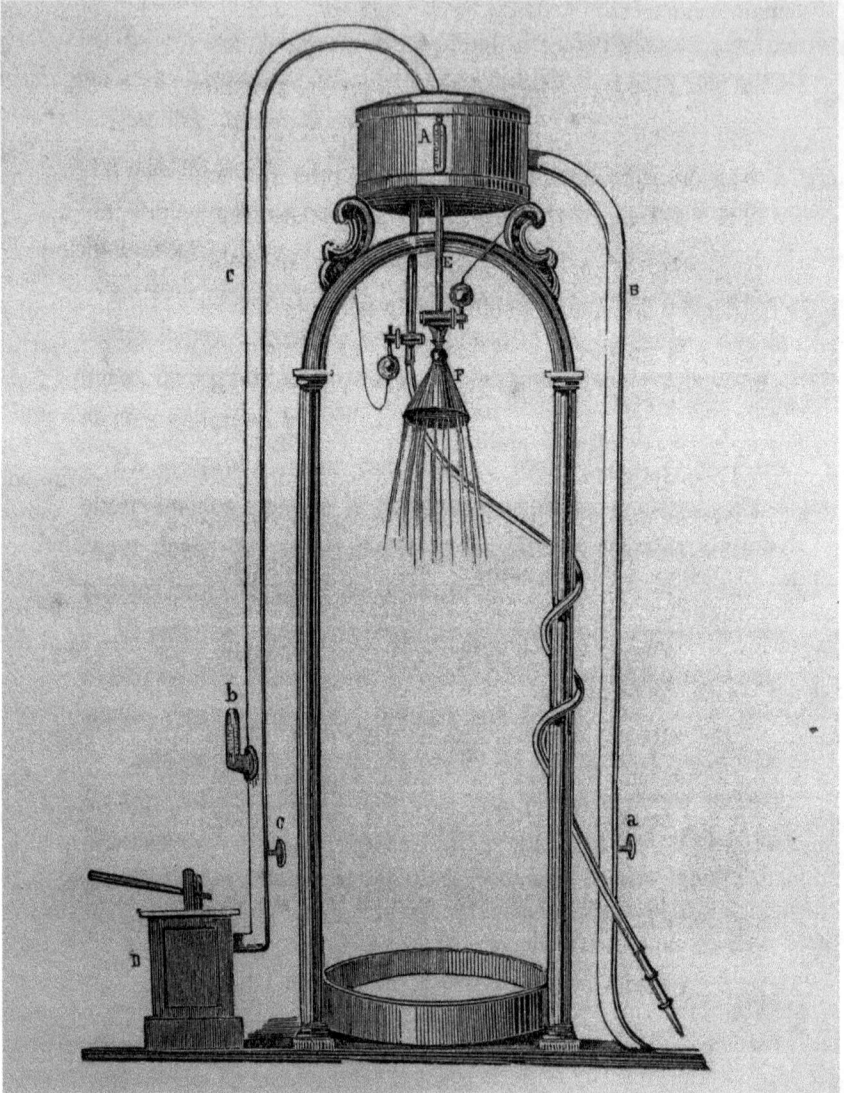

FIGURE 8. "Shower, Fleury Design." Lugo 1875: 17.

Mexico City bathhouses installed showers modeled on Louis Fleury's equipment in his baths at Bellevue-chez-Meudon, France.[42] Therapeutic bathing was further institutionalized in the National School of Medicine, where numerous theses were produced on the topic between 1875 and 1910. All of these hydrotherapeutic showering and bathing facilities utilized the waters of the public water system, irrespective of their origin in the Chapultepec springs, the springs of the Desierto de los

Leónes, the new *pozos artesianos*, or the Xochimilco springs. In these versions of hydrotherapy it was the physical force of the water, more than its mineral content, that was considered curative.

Interest in the therapeutic qualities of mineral springs never died. The medical history of mineral waters in Mexico has been ignored by historians who focus instead on the botanical elements of medicine, but we know that scientists and doctors did work on these extraordinary waters.[43] The Royal Botanical Expedition to Mexico in the 1790s carried out studies of the waters of Cuincho and San Bartolomé, and Jardín Botánico director Vicente Cervantes extracted the minerals from them by evaporation.[44] In 1795 Antonio De la Cal y Bracho, who took Cervantes's 1792 course on *botánica* and became the correspondent in Puebla for the Real Expedición, carried out an identical study of the mineral waters and local plants of Tehuacán.[45] Although he did not further explore the place of the *"reino mineral"* in Mexico's national popular-medicine tradition, focusing instead on *botánica,* he clearly defined the need for further studies of mineral springs.[46]

In 1850 most hot springs in Mexico were rustic and undeveloped, and the bathhouses that did exist dated to the colonial era. Regardless, people in Mexico continued to utilize hot springs for bathing, drinking, and even inhaling cures. In 1844 Ernesto Masson lambasted the 1790 remodel of Peñón, saying that "everything about the place reveals the poor taste of the era."[47] But this was also his dismissal of the everyday folks who kept the baths "in vogue" throughout the years, and their "vulgar" ideas about the medicinal properties of the waters. The humble used the bathhouse, and the very poor simply took half-baths sitting in the drainage canal outside of the bathhouse.[48] Masson notes that while the Mexican elite was developing an interest in scientific therapeutic bathing, this did not result in improvements in Peñón and greater use of the waters. The rebuilding of Peñón may have been frustrated by the inheritance dispute discussed in chapter 3, but elites did not travel, as their European counterparts did massively throughout the nineteenth century, to any other hot springs until around 1890, when hot springs bathhouses were built in Aguascalientes, Tehuacán, and Topo Chico.

Those who sought out and wrote about Mexican mineral springs were often foreigners who had experience with European health spas. In 1835 Francesco Antomarchi, Napoléon Bonaparte's last doctor in Corsica, lived for a brief period in Mexico and visited a number of hot springs: Xochitepec (Morelos), Atotonilco de Santa Cruz (Zacatecas), Ojocaliente (San Luis Potosí), and "Agua de San Ramón" (Aguascalientes). Antomarchi, who would die in Santiago, Cuba, in 1838, was ill at that time, and it is likely that he was searching for a cure. Just a day's travel south of Mexico City, the Atotonilco springs were hemmed by a masonry wall that formed a pool where men and women bathed together. In Ojocaliente there were two well-kept bathhouses attached to the pool, one for each sex. Antomarchi conducted the customary analysis of the waters of each—temperature, chemical content—and derived a determination of their usefulness for treating different medical conditions.[49] Regino Gayuca, who transcribed this report in 1843 for the journal of the National Museum of Mexico, wondered why "if in many parts of Europe they value thermal waters and have identified their minerals, we hardly mention those that we have in this America?"[50] Aguascalientes had been known for its curative hot springs since its foundation in the sixteenth century, and Antomarchi, together with a group of local intellectuals, conducted an analysis of the waters of the San Ramón springs.[51] None of these springs attracted the interest of cosmopolitan urbanites, and none were developed into bathing establishments until the 1880s.

Any mention of hot springs by proponents of therapeutic bathing was inevitably followed by dejected comparisons to the advanced state of installations and practices elsewhere. In 1840 Francis Erskine (Fanny) Calderón de Vaca visited Peñón de los Baños, told of its decrepit state, and offered a vision of the thriving business that could be built at the site by an "enterprising Yankee." Between 1844 and 1849 Ernesto Masson, a naturalized French immigrant, carried on a heated debate in Mexico City's press with the goal of improving the state of the Peñón baths, which was mired in an inheritance dispute, so that patients could make use of its "astonishing virtues."[52] He lobbied the city government to expropriate Peñón and sell it to Anselmo Zurutuza, who promised to build "a European-style thermal bathhouse."[53] Ramón Malo, governor of the Distrito Federal (DF), ordered the National Health Council to study Peñón and identify its medical benefits, in the model of countless other hot springs studies, and Leopoldo Río de la Loza and Ernesto Craveri published their report in 1849. By then, however, the DF had a new governor, Pedro Jorrín, who did not care enough about Peñón de los Baños to proceed further. In 1858 a similar study was carried out in the mineral springs of Tehuacán,[54] and Peñón's waters were analyzed again a few decades later in Paris,[55] but the springs and their ancient bathhouses did not receive investment by developers.

Mexico's pharmacists, however, did make a business of mineral waters. Whereas the 1846 *Farmacopea Mexicana* included water as a medium for preparing herbal infusions (*agua de azahar, agua de canela, agua de hinojo,* etc.), the *Nueva Farmacopea Mexicana* published in 1874 included an appendix on "Waters" with chapters on potable waters, natural mineral waters, and artificial mineral waters, the last with recipes for the contents of famous European mineral springs. Included in this appendix are chemical analyses by Río de la Loza of various Mexican hot springs, as well as by Baguerisse and Lambert. Mineral waters had become part of Mexican modernizing medicine, and Mexican doctors and pharmacists were influenced by the growing role of hot mineral waters in European medicine. For example, Plácido Díaz's 1876 study of the hot springs in Puebla pointed to their usefulness in treating tuberculosis, an analysis that built on the European tradition of treating that disease with fresh air and mineral springs.[56] In 1878 the Mexican National Academy of Medicine announced a competition to study Mexico's mineral water, indicating the resurgence of mineral waters in medicine at the time.

José Lobato won the contest with a comparative analysis of the springs in Villa de Guadalupe and Peñón de los Baños.[57] The competition was one of two about water that were announced in 1874: Antonio Peñafiel won the other with his treatise on potable waters.[58]

Lobato announced the victory of scientific medicine over "vulgar," "empirical" traditions of hydrotherapy, writing that "little by little the belief in the therapeutic effects of mineral waters has turned into a scientific doctrine, and that this has become known to all social classes in the civilized countries of Europe, America, Asia, etc."[59] Lobato's analysis and classification of the mineral waters was based in European models, but he adjusted them to grapple with the specificities of the mineral waters in Mexico. In doing this Lobato built upon the tradition of studying the chemical and mineral qualities of water that was developed by Río de la Loza. He privileged the geological origins of the waters in his classification, complementing the therapeutic orderings proposed by French scholars and doctors Etienne Ossian Henry, Maxime Durand-Fardel, and Jules Lefort. Lobato established seven families, fourteen classes, fifty-seven genders, and a scattering of species of mineral waters in his system, all according to their chemical composition and geological origins.[60] The therapeutic agency of a water, described in the eighteenth century as its "virtue," was recast as a "medicinal, mineralogical principle that gives expression to a medical power."[61] This science did not rid waters of their efficacy.

STRUGGLING TOWARD SPAS

In the 1870s, mineral springs in the Valley of Mexico were converted into bathhouses, part of the wider profusion of bathing at the time. The hot springs bathhouse of Peñón de los Baños languished in a rudimentary state, but bathhouses were built at two sources of ferruginous (iron-bearing; also known as "chalybeate") waters to the north of the historic center. The spring at Aragón was located on the side of the road leading into the religious center of Guadalupe. The owner of the land was prompted by the rising popularity of bathing in the 1870s to unearth the spring, which until then had been considered a nuisance. He commissioned an analysis of the waters by the chemist Gumersindo Mendoza, and built a bathhouse in 1875 with a few *placeres* in small private rooms, a garden, and a ten-by-ten-meter swimming pool.[62] Soon a steady stream of patients treated anemia and other maladies with the iron-rich waters. The other mineral water baths near Guadalupe were named the "La Estación," and were located a few steps from the station of the train that brought visitors from Mexico. Eduardo Liceaga built that bathhouse in 1878, supplying it with an artesian well perforated by the Beléndez and Velázquez company. It had six "first class" rooms with *placeres,* a bottling room, a gynecological treatment room, a room of showers, a garden and more, and was designed in the neoclassical style of a Pompeian villa similar to that of the bathhouse in the Bosque de Chapultepec.

FIGURE 9. El Pocito, Villa de Guadalupe. Michaud 1874. With permission of Universidad Nacional Autónoma de México, Instituto de Investigaciones Estéticas, Archivo Fotográfico Manuel Toussaint, Colección Julio Michaud.

These bathhouses were built with clear scientific medical justifications. Nevertheless, they both served a clientele that was drawn to Guadalupe by the religious fame of the mineral waters of the nearby Pocito, a spring that became a Christian holy site in the sixteenth century and was considered to mark the site of the appearance in 1531 of the Virgin de Guadalupe to Juan Diego. People considered the waters to be "miraculous" and curative, and the crowds that came to drink and bathe were so great that the church erected a structure around the spring in 1648, and in the 1770s built a baroque chapel (see figures 9 and 24). The spring welled up in a two-meter receptacle inside the chapel, with a grate on it that prevented people from climbing in to bathe. A copper cup on a chain was attached to the grate so that pilgrims could drink the waters. Río de la Loza analyzed the waters of the Pocito and argued that they were more effective than similar springs in Europe, and far better than the patent medicines being produced at the time.[63] He found the waters of La Estación and Aragón to be similar but not identical in their mineral composition.

Lobato judged most Mexican bathhouses woefully underdeveloped in comparison to the spas of Europe. He decried the failure to institute medical hydrotherapy in terms of science and tradition, but was aware that this had a lot to do with social class. A major obstacle to the development of modern bourgeois bathing was the "routine of tradition" that governed the therapeutic use of these waters. Most bathhouses served humble clients who had, for centuries, made the pilgrimage to Guadalupe for the miraculous properties of the waters, but who could not afford, nor were interested in, expensive and lavish facilities. Bathhouses, Lobato argued, should be run by doctors and trained managers, much like textile factories should be run by directors and mechanics.[64] These doctors and managers should be schooled in the latest science in order to deal with the complexity of the mineral content and temperatures of the waters and their application to a variety of maladies through a variety of systems: tubs, showers, inhalation devices, etc. Empiricism and routine—the heart of popular medical traditions—prevented the implementation of more sophisticated and effective mineral water therapies.

Both business and government needed to intervene, Lobato argued. The two bathhouses in Guadalupe were attracting visitors from Mexico City, but neither was adequately capitalized, and hotels, guesthouses, doctors' offices, restaurants, and other amenities common to the European spa towns such as Vichy were completely lacking. While the business of bathing might not be rewarding, baths were also a service offered to the public, and the meagerness of profits should not deter Mexican investors from building first-rate spas where scientific medicine could flourish. Government needed to help as well. Mexico City's Distrito Federal had no medical inspector to oversee hydrotherapy treatments; there were no doctors attending to the sick at the bathhouses. The Consejo de Salubridad, Lobato argued, needed to treat mineral waters as a public health issue. "In our country," he stated, "we still today have no knowledge of the regulations of sanitary police that are required for bathing establishments of this kind."[65]

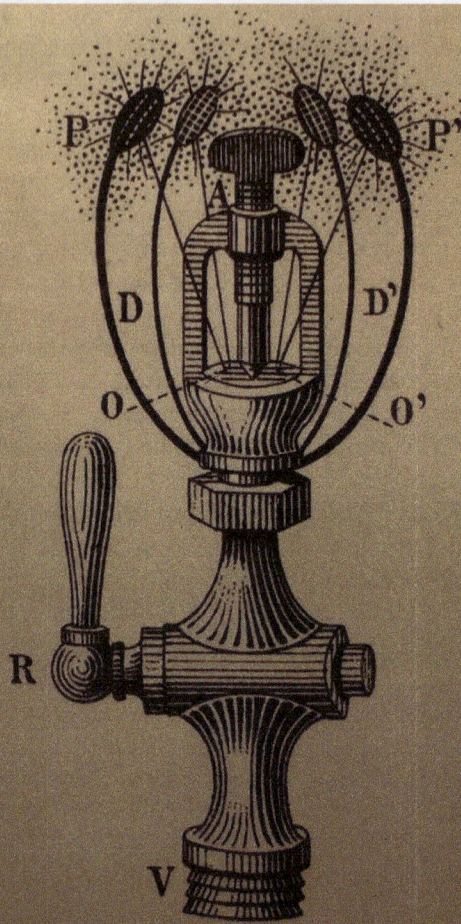

FIGURA 7.ª

Aparato de pulverizacion para pulverizar
las aguas minerales en los departamentos
balneatorios de inhalacion.

FIGURE 10. "Misting apparatus for pulverizing mineral waters in the inhalation departments of the baths." Lobato 1884: insert between pp. 192 and 193.

While in the late eighteenth century the ruling class policed the sexuality and sociality of plebian bathing, for Lobato policing was needed primarily to counteract the weight of popular medical traditions: empiricism, folk knowledge, religious beliefs. "Take a look at the buildings in Aragón and Guadalupe, on the one hand,

and Peñón, on the other, and you will see that they are none other than common baths, fitting for a population that has little civilization, scientifically and socially speaking."[66] There was the lack of mass appeal for bourgeois scientific bathing practices and bathhouses, and crowds still administered their own treatments at Peñón, despite its decrepit state, because of the widely held idea that they were useful for treating rheumatism and infertility among women. Lobato dismissed the plebeian bathing tradition as superstitious "empiricism" that eroded the prestige of those hot springs among scientists. The mass appeal of the Pocito de Guadalupe was due, he argued, to the fact that "the Spaniards made indigenous converts believe that the spring is miraculous and supernatural."[67] So strong was the belief in the holy, curative powers of the waters that faithful craftsmen and even elite matriarchs donated their Sundays to building the chapel in the 1770s—no labor was hired.[68] And this was the root of the issue: the reforming mineral water doctor was promoting a water culture that was not shared by almost anybody else in Mexico.

"Mexican mineral hydrotherapy is destined to figure notably in the annals of science."

—José Lobato, 1884

Source: Lobato 1884: ix.

Porfirian scientists often complained about a lack of basic science concerning mineral waters, but by 1886 there were already analyses of 116 different Mexican mineral springs.[69] The same impression of paucity was shared by French colleague Emile Delacroix, whose 1876 study of the world's mineral waters mentioned Peñón and Guadalupe, but found that the curative properties of their waters were yet to be properly classified.[70] Clearly there were long-standing traditions of taking the waters among everyday Mexicans, but in the early 1880s Mexico did not have a robust enough bourgeoisie with modern Europeanized concepts of health, medicine, and leisure to support investment in mineral water spa establishments such as those in France and Germany. As Friedrich Semeleder, former doctor to Emperor Maximiliano and Empress Carlota and a member of the Mexican Academy of Medicine, put it, "in Mexico there is not the same kind of mania for mineral waters that there is in Europe."[71]

Some doctors and politicians rejected the Eurocentrism of spa medicine, framing their project instead in terms of the specificity of Mexican bodies and environments, a nationalist, climatological strain of thought with precursors in the late eighteenth century. In 1886 the National Academy of Medicine, "mortified" by the idea that Europeans would hold more interest in their hot springs then they did, approved a national-level study of Mexico's mineral waters with the purpose of generating scientific information to support medical applications and the development of spas.[72] In 1889, Carlos Pacheco, Minister of Development, created the National Institute of Medicine (NIM), with the mandate to "possess truths discovered in this country, and perhaps in some cases only applicable to this country."[73] At the same time a major survey was conducted by the Ministry of Development to collect climatological information from the 2,863 municipalities of Mexico, an interest that was shared by Liceaga.[74]

> "If, because of our climate, geography, race and customs we have a different physiology, idiosyncrasy, morbid receptivity, and constitution; if our fauna, our flora and our waters are not the fauna, flora and waters of other places: why, then, if we have such varied national elements, have we not created a national science?"
> — Secundino Sosa, 1889
>
> Source: Sosa 1889a: 2.

Pacheco believed in the therapeutic efficacy of water. He was a regular visitor to the Alberca Pane, where he swam in the pool and used the baths.[75] Hydrotherapy was a central focus of climatological models of health,[76] and Pacheco had a special interest in developing both the science and business of mineral springs, which were, in the words of Secundino Sosa, director and founder of the NIM's journal *El Estudio,* "almost completely abandoned."[77] Over the next two decades the NIM conducted an ongoing effort to study the country's waters and "form a hydrological repertoire with chemical and therapeutic uses."[78] Whereas Lobato called on government and business to develop Mexico's mineral waters into spas, Sosa argued that it was the doctors who had to bring together the science of hydrotherapy, the capital needed to build spas, and the clients to keep those spas functioning.

Eduardo Liceaga was the kind of doctor that Sosa was interested in; one who promoted the science as well as the business of bathing in Mexico. He was keenly interested in the role of water in public health, and pushed therapeutic bathing with the same conviction that he promoted modern water supply systems and sewers. Liceaga was born in Guanajuato in 1839 into a family of doctors, and graduated from the National School of Medicine with honors in 1866; Leopoldo Río de la Loza was a member of his exam committee. He was at the center of the worldwide turn to microbiology, visiting the Pasteur laboratory in the 1880s and returning to Mexico with plans for inoculations against rabies. He served as president of the Consejo Superior de Salubridad and director of the National School of Medicine and used these positions to elaborate building and sanitation codes for Mexico City. He also served twice as president of Mexico's National Medical Association and oversaw the construction of the General Hospital.[79] Liceaga collaborated with

Roberto Gayol, a hydraulic engineer who served as assistant director of public works for the Mexico City government and sat on the Consejo de Salubridad Pública. Together they designed and built the General Hospital, as well as the city's drainage system. The General Hospital, initiated in 1896 and concluded in 1905, was equipped with a hydrotherapy building that offered a variety of medicinal, therapeutic, and hygienic encounters with water.[80] There was a swimming pool, Russian baths, Turkish baths, all kinds of showers, nozzles and sprayers for therapeutic applications, and cold and warm showers for personal cleanliness modeled after those used by the French military.[81] The complex was supplied by artesian wells as well as city water and had a robust sewer system.

"That was precisely what [the girls'] mother and Dr. Liceaga sought. By enhancing the body's circulation, hydrotherapy bestows on the nervous system—which is so delicate, so exquisite, and so obedient—a far from negligible amount of what can be called the joy of living. . ."

—José de Cuéllar, 1941

Source: De Cuéllar 1941: 23–24.

Doctor Liceaga was also deeply involved in the business of therapeutic bathing. He built "La Estación" bathhouse in 1878 and in 1880 published a study of the different springwaters of the Villa de Guadalupe.[82] A decade later he turned his attention to a study of the mineral waters of Peñón, commissioned by Manuel Romero Rubio, secretary of Gobernación and father-in-law of Porfirio Díaz.[83] Between 1887 and 1892 Romero built a sumptuous, modern spa at Peñón and Liceaga's study was, like all studies of mineral waters, both science and promotion. Liceaga presented it to the Mexican National Academy of Medicine, and immediately had it translated for distribution at the 1892 meeting of the American Public Health Association that he organized in Mexico City.[84] Doctor Friedrich Semeleder described the study as "advantageous even from a financial point of view."[85]

The new Peñón bathhouse was as close to a European spa that could be found in Mexico. The bathhouse itself had two floors, the lower floor with a bathing area for men, decorated in Egyptian motifs, and one for women in an Aztec style. This mix of decorative elements from Old World and New World civilizations evoked the classical, Mediterranean roots of bathing in Mexico, participated in the Egypt-mania of the time, and by claiming a classical tradition of bathing was an assertion of Mexico's place among the world's civilized countries.[86] The bathing apartments each had a bathtub room and another room where guests could recline on a bed and sweat. There was a room of showers, a sauna, and fountains of the mineral water for drinking. On the upper floors there were sumptuously furnished

FIGURE 11. Departamento de Baños, Peñón de los Baños bathhouse. *El Mundo Ilustrado* 2, no. 12 (1906). With permission of Universidad Nacional Autónoma de México, Hemeroteca Nacional de México, Fondo Reservado.

bedrooms and meeting rooms, and nearby buildings held a chapel, a manager's quarters, a billiard saloon, a restaurant, and a bowling alley. The building offered "the most beautiful views of the Valley of Mexico" to aid the rest and recuperation of the clients.[87] President Díaz himself reserved a special suite of rooms for his family at the spa.[88]

The new bathhouse was directed primarily at the Porfirian bourgeoisie. Different "classes and prices" of baths were available, allowing some of the humble folks who had used the waters before the bathhouse was built to continue to do so, but poor people no longer had access to the used waters in the exit channel. Liceaga provided the bourgeois clients with a guide to the rules of behavior at modern spas, so that the curing properties of the waters were complemented by diet, hygiene, rest, diversion, and leisure. Patients were advised to "change their habits entirely," to leave the stuffy houses and offices behind along with all the excitement and worries of business, rich food, late nights, alcohol, and "theaters and balls" that the bourgeoisie was accustomed to. He warned that treatment required prolonged stays, repeated over many years, and gave more precise instructions about bathing and drinking to address particular maladies.[89]

FIGURE 12. "Edificios de 'El Peñón.'" *El Mundo Ilustrado* 2, no. 12 (1906). With permission of Universidad Nacional Autónoma de México, Hemeroteca Nacional de México, Fondo Reservado.

At 11:00 a.m. on November 31, 1892, a caravan of horse-drawn coaches rolled up in a salty, dusty cloud before the sparkling Victorian buildings at Peñón de los Baños. Two hundred formally dressed men, most of them doctors, made their way into the brand-new bathhouse and bottling plant situated at the foot of a rocky outcropping in the dry bed of Lake Texcoco. The physicians were from Mexico and the United States, and conversed in Spanish, English, and French. They were met by a cadre of some of the most influential figures in Mexico, including the owner of the installations, President Porfirio Díaz's father-in-law, Manuel Romero Rubio, and Díaz's personal doctor, Eduardo Liceaga. The foreigners were deeply impressed to find such a refined and sumptuous establishment, reminiscent of the luxurious spas of Europe, on the forlorn outskirts of Mexico City. They reviewed the marble baths, examined the hygienic bottling plant, and strolled through the richly appointed hotel before making their way back to the coaches for the seven-kilometer ride to the Castillo de Chapultepec, where they were received in the Yellow Room by Liceaga and the president of the Republic himself.

Source: Liceaga 1892: 3, "Report" 1893, "American Public Health Association."

Peñón was one of a small number of hot springs spas built in Mexico during the Porfiriato. After 1884, the railroad that passed through Aguascalientes brought visitors to its new bathhouses, and the Topo Chico bathhouse near Monterrey was erected in the 1890s, together with a solid and elegant hotel (see chapter 7).

Porfirio Díaz ordered studies conducted of the springs at Tehuacán, which led to the development of the Balneario del Riego and other bathhouses.[90] As we have seen in chapter 5, taking the waters in mineral springs was part of a wider upsurge in bathing and swimming fueled by the perception of hydraulic opulence associated with the artesian well. Bathhouses and mineral springs resorts multiplied over the next fifty years, but hygienic bathing slowly shifted from a public, social activity to a private, individual one, carried out increasingly within the confines of people's homes. Mineral waters bathing remained a social activity, but lost much of its medical rationale as therapy gave way to leisure as the principal rationale of the business of bathing.

FROM BATH TO SHOWER

"Hygiene and therapy fight for dominance in hydrology," Secundino Sosa announced in 1889.[91] In the nineteenth century, chemistry, long the protagonist in the science of water and public health, slowly ceded ground to biological views of the importance of microorganisms, marking a similar movement away from treating bodies by exposing them to minerals in waters, and toward protecting bodies from the bacteria in the liquid. Liceaga, Lobato, Sosa, and many other Porfirian doctors held a therapeutic understanding of the efficacy of waters that was based in chemistry, and sought to promote and profit from European traditions of mineral springs bathing. But the new attention to microbiology was changing people's relations to water, and to each other. Instead of a medium for mineral treatments, water increasingly came to be seen as a carrier of contagious microbes. The contradiction between cure and contagion was sharpened by the idea that bathing could also promote hygiene and health by washing away biological contaminants.

The shower, taken individually, was seized upon as the way to benefit from the cleansing effect of water while ensuring that the water did not move microbes from one body to another. Showers produced a constant circulation of water, and like flushing, appealed to sanitarians preoccupied with stagnation. During the Porfiriato they were increasingly viewed as the most modern, healthful form of bathing. In 1885, a traveler in Mexico commented that the "bathhouses with showers, already common in Mexico, are challenging the bathhouses with tubs and *placeres* that are so common in this country, where all social classes frequently bathe."[92] With showers in the public baths, the scene described with horror by Antonio García Cubas of a poor mother bathing her entire family in the same tub of water would be erased.[93]

The resurgent field of hydrotherapy also promoted the displacement of bathtubs by showers. Hydrotherapy held that it was the physical action of water on the body, rather than the minerals of chemicals contained by the water, that was therapeutic. This viewpoint gained strength from the knowledge that minerals suspended in water were actually not absorbed by the skin. In Mexico, popular hydrotherapy, or hydropathy, had been rejected by many doctors as an empirical, unscientific practice in the 1840s, but medical students continued to submit theses on "scientific hydrotherapy" for the professional examination at the National School of Medicine throughout the last half of the century.[94] The "heroic method of cold baths" inherited from the hydropathy of Priessnitz and others enjoyed a resurgence between 1890 and 1910 in many countries; first Germany and Austria, then the United States, Mexico, and elsewhere. Columbia University even dedicated a faculty chair to hydrotherapy, the first in the hemisphere.[95] Hydrotherapy was applied by graduates of the National School of Medicine in the new General Hospital in Mexico City.[96] Part of the success of this medical tradition came from its ability to incorporate the latest scientific discoveries, such as electricity and radiation.[97] In 1901, for example, Samuel Morales opened an ultra-modern "electro-medical" bathhouse in Mexico City, offering shock treatments in cold and hot water baths, dry-shocks, x-rays and massages.[98]

Following the initial inspiration of Priessnitz in the 1840s, and Fleury in the 1870s (Lugo 1875), hydrotherapists elaborated an increasingly complex array of hoses and showers to direct water at particular parts of the body. By 1900, such showers were present in hospitals and baths around the world—the Orthopedic Hospital in Philadelphia; Elizabeth Hospital in Washington, DC; Massachusetts State Hospital in Danvers; and the Riverside Baths in New York City—and were installed by Eduardo Liceaga at the new General Hospital in Mexico City (1905). Liceaga also promoted hydrotherapy from his chair at Mexico's National School of Medicine.[99] The psychiatric hospital "La Castañeda," opened in 1910, deployed hydrotherapeutic showers to treat mental and emotional disorders, as did Bellevue Psychiatric Hospital in New York.

The turn to the hydrotherapeutic shower was a turn away from the bathhouse culture of the late nineteenth century. Victor Macías-González (2012) shows that Porfirian bathhouses contained varied social spaces that offered a wide range of leisure activities, from swimming pools, barbers, and massages to reading rooms and restaurants, as well as numerous private rooms that facilitated homoerotic encounters among middle class and elite Mexican men. These bathing establishments often embraced an imaginary of opulence and leisure associated with Roman and Turkish baths. But in the face of growing homophobia propelled by the notorious persecution of the "41" in 1901, bathhouse operators such as the owner of the San Felipe Baths stressed "order and morality," limiting physical and social interactions and focusing activity on the hygienic act of washing quickly in strictly individualized spaces. The revolution (1910–20) added to this tendency with a critique of the decadence of the Porfirian bourgeoisie, and the promotion of ideals of efficiency, action, and heterosexual virility. Revolutionary reformers in the Departamento de Salubridad lauded the shower as a fast, efficient way to wash the body that eliminated the sensuality and connotations of homoeroticism carried by *placeres* and other immersion baths.[100] In the twentieth century, the policing of sociality and sexuality in the bath compelled a transition from immersion to showering.

The extension of hydraulic infrastructure into the household generalized the shower and individualized bathing. Widespread everyday bathing only became possible with the opulence of water that began with the artesian wells after 1850. Bathhouses grew in popularity in the last decades of the nineteenth century, but this expansion of social bathing was matched after 1910 by private bathing as huge volumes of new water were delivered by the aqueduct from the springs of Xochimilco to homes in the growing suburbs along the Paseo de la Reforma. The Colonia Doctores, Colonia Roma, Colonia Juárez, and Colonia Condesa were obliged by the 1891 Sanitary Code (passed by Eduardo Liceaga when he directed the Consejo de Salubridad) to include sewers and potable water lines, and all new housing was to be delivered water individually.[101] People would no longer rely on collective fountains and wells that had served the city's inhabitants for centuries.

The shower was looked upon by sanitary reformers as the most progressive and modern mode of bathing and the "most commonly used in the civilized countries."[102] It was hygienic, therapeutic, and allowed the bather to adjust the water temperature, thus eliminating interactions with bathhouse workers and servants. In the 1920s the Departamento de Salubridad passed Sanitary Engineering and Potable Water regulations that required that each apartment in a building or a *vecindad* have an individual water meter, and that showers be installed in all private housing, new and old.[103] In the draft of the Regulations for Public Baths written in 1924, the Departamento de Salubridad required that public pools be emptied, washed, and refilled with new water twice a week, and swimmers were to shower before entering the pool. Public baths were obliged to provide sponges, soaps, and other implements to individual bathers and these items were to be disposable. Bathhouses dedicated to hygienic rather than medicinal bathing were required to substitute the *placer* or tub with showers in private rooms.[104] In medicinal, mineral springs bathhouses such as Aragón and Peñón, the Departamento de Salubridad allowed bathing in tubs, but required them to be cleaned daily and forbid the use of bathwater by more than one person.[105]

The business of bathing changed over the first few decades of the twentieth century, as more wealthy and middle-class Mexicans took showers at home, abandoning the bathhouses to a swelling urban underclass. Bathhouse owners argued that Mexico's bathhouses were already better than those of New York, Chicago, Paris, and London, and that the additional expenses of installing showers were unnecessary and would make bathing too costly for the "the middle and humble class, employees and workers who form the great majority of the users."[106] The Alberca Pane, the once-thriving establishment where Carlos Pacheco, Porfirio Díaz, and a host of other elite customers met to swim, soak, steam, and socialize, was apparently unable to pay for the required remodeling, and asked for an exemption from the new rule to install showers.[107]

Salubridad's effort to reshape bathing ran into opposition from owners of baths, real estate developers, and landlords. The department convened a meeting between government officials, property owners and developers, and sanitary engineer Roberto Gayol. The young Salubridad officials insisted that the landlords and developers be required to install a shower for every dwelling, and one shower for every twenty people living in collective dwellings such as *vecindades* and old houses that had been divided into apartments. The property owners protested that costs would be prohibitive, and that such measures were wasted on a public that was not particularly clean. Salubridad countered that showers were needed to create new habits of hygiene among the poor, but that bathing itself did not need promotion: the bathhouses were thronged, more than ever before, by working-class and poor city dwellers. Rather, the new habit they sought to promote was showering at home, and it was indeed adopted en masse during the twentieth century, displacing public bathhouses almost entirely.

CONCLUSIONS

Roberto Gayol eased into his chair at the new headquarters of Public Health, built on the Paseo de la Reforma at the edge of the Bosque de Chapultepec, and admired the modern art-deco architecture. Gayol remembered when this part of the city was still open fields, when he directed the construction of water and sewer lines that serviced the new middle-class neighborhoods erected after 1900. All those houses had running water and bathrooms, and now the young engineers of Public Health were attempting to bring these amenities to the rest of the folks in Mexico City. He admired their revolutionary zeal to build a new, more integrated nation, but had his doubts. The Sanitary Engineering Code they presented at the meeting made showers obligatory in all dwellings, but the landlords and developers at the table fought the measure. Gayol agreed with the landlords that it would be too expensive and difficult to retrofit the colonial period buildings in the city center, but he did not second their opinion that poor Mexicans who lived in older sections of the city in *vecindades* [tenement blocks] and apartments without plumbing "were just plain dirty." Showers should be available to those people too, and he agreed with the *licenciados* of Public Health that the infrastructure would promote good habits of hygiene. At 73, he had lived long enough to recognize that the revolutionary effort to create a culture of bathing around household showers was the logical conclusion of the modernizing hydraulic engineering project that he had begun fifty years earlier.

Source: AHSS, FSP, SSJ, Caja 21, Exp. 9, Transcript of Meeting (April 9, 1930).

After 1850, bathing changed in important ways, a result of hydraulic opulence, capital looking for profits, and new scientific knowledge of chemistry and microbiology. The idea of water moved toward that of a uniform substance, delivered through encompassing infrastructures. But at the same time, everybody recognized a diversity of tastes and qualities among the different water sources that supplied Mexico City. By 1858 the hard water of the Chapultepec springs and the soft water of the aqueduct from the Santa Fe springs were supplemented by wells in Bucareli, Los Migueles, and the Calle de Cordobanes. These waters and others were analyzed by chemists who described their temperature, density and levels of oxygen, carbonic acid, calcium sulfate, bicarbonate of calcium, and other contents.[108] Waters had always been recognized for their various qualities (*gorda, delgada, gruesa, dulce, salada, hedionda, azufrosa,* etc.), and these distinctions could increasingly be explained by chemists in terms of content of minerals, gases, organic matter, and the like. Mexican chemists, like their European counterparts, understood that different waters had different uses: "some were destined to satisfy household needs, others for industrial ones, and not a few for restoring the health of man."[109] Some waters, such as those from the artesian wells near San Lázaro, were unpalatable and smelled badly because of dissolved gases, and were not considered useful. In an interesting paradox, waters continued to be viewed as plural and specific because scientists and planners sought to combine them and convert them into a singular substance.

The sciences of chemistry, microbiology, and medicine grappled with the specificity of all these waters. The discovery by Pasteur and his contemporaries of the microscopic animals responsible for fermentation as well as sickness placed medicine on a new footing, as health was resignified as cleanliness, and cleaning made into a battle against germs. The hegemony of microbiology and hygiene rebalanced the range of acceptable purposes for bathing, but it was a fragile hegemony. Water used for cleanliness was also a threat to hygiene, as it could just as easily bring bodies into contact with those elements that were considered dangerous. Bathtubs and soaking gave way to showers and rinsing, ensuring that once washed from the skin, "dirtiness" was banished rapidly down the drain. Newly perceived dangers of baths came to compete with long-held ideas about the therapeutic benefits of water, and the cleansing flow of the shower stood out as the most hygienic interaction with the liquid.

Despite the rise of the hygienic understanding of water, the therapeutic, ludic, and recreational dimensions of bathing were not displaced by the urge to clean, but rather flourished as a parallel set of activities. The long shift to the shower and the rise of the singular concept of sanitized water did not eradicate the engagement with diverse waters. The increased flow of capital toward the business of bathing was accompanied by an evolving discourse and practice of water therapy that swirled at the fluid edge of medical orthodoxy. Bathhouses in the late nineteenth and early twentieth centuries such as the Alberca Pane offered an ever-wider array of bathing experiences: hot, lukewarm, and cold waters, plunge baths, steam, dry saunas, *placeres,* swimming pools, inhalation chambers, drinking fountains, and showers of all different kinds. After 1920 the solitary household shower with public water grew to be the most important daily contact most city dwellers had with the liquid, but heterogeneous waters and water cultures lived on in the hot springs resorts and bottling plants that flourished throughout Mexico during the twentieth century.

Dispossession and Bottling after
the Revolution

Mineral springs attracted renewed attention during the late nineteenth and early twentieth centuries. Artesian wells and expanding urban infrastructure produced an opulence of public water that enabled bathhouses to proliferate, reshaping bathing practices and socialities. At the same time, however, the homogenization of water actually boosted the value of heterogeneous waters, and investors turned to mineral springs with newfound interest. During the Porfiriato, spas and bathhouses were built and rebuilt in Guadalupe and Peñón, and new railroads brought tourists to baths in Aguascalientes, Topo Chico, and Tehuacán. In addition to the expanded bathhouses, entrepreneurs took advantage of widespread and long-standing ideas about the curative efficacy of mineral waters and opened bottling plants at those sites.

Historian Luis Aboites (1998) describes a long process between 1880 and 1946, facilitated in some ways by the Mexican Revolution, of expanding national-state control over water resources. The narrative of state centralization certainly captures much of Mexican water history, but I wish to narrow the focus on the mechanisms by which centralization proceeded. The history of the Tehuacán and Topo Chico mineral springs reveals the ultimate beneficiary of state centralization to be the private bottling industry, a process that might be more accurately called primitive accumulation. Karl Marx depicted this process as a violent rupture of customary property relations "written in the annals of history in letters of blood and fire."[1] In twentieth-century Mexico, however, the dispossession of water resources was more often realized through legal and political mechanisms supported by technoscience, and the emergent fields of hydrology and hydraulic engineering were particularly important.[2] This was a quotidian, cultural process involving the authority of certain kinds of argument, reasoning, and evidence, and an engagement with the bureaucratic procedures of the state. Also, while these waters were bottled for exchange in the marketplace, it was the assumption of heterogeneity and singularity—their culturally formed use-values—that drove their commoditization.[3]

THE BUSINESS OF BOTTLING MINERAL WATERS

Since the Middle Ages in Europe, pilgrims who could not make the long trip to springs such as Lourdes were still able to secure the effects by quaffing bottled water. Bottled mineral waters became increasingly common in the early nineteenth century in Europe and North America. Part of this was due to the expansion of transportation infrastructure that made it much cheaper to bring the curative waters to their hopeful consumers in the cities. Bottling was a business venture that was both profitable and promoted public health, yet did not require patients to visit the bathhouse. "The exportation of these waters great distances within the country is very easy," José Lobato pointed out in 1884, "and should be as beneficial to the sick people who are treated with this medicinal water by ingestion, as it is to the bottler who knows how to set up shops in the capital of every state."[4] Another reason for this growth in bottling was that the local water sources in the growing cities were increasingly contaminated, and mineral waters bottled at faraway springs were less prone to contamination. Notions of the therapeutic character of diverse waters, as well as the unhealthfulness of the homogeneous water that was served through public pipes, bolstered the value of heterogeneous waters.

Ideas about why waters were curative, and how to administer them, increasingly favored bottling. By the 1880s doctors had concluded that the skin was an effective barrier to the absorption of minerals from water, and that the minerals in the waters needed to be delivered through inhalation and ingestion rather than bathing. In his study of mineral waters in Mexico, José Lobato describes inhalation techniques, pioneered in Germany, that required water to be "pulverized" into a mist(see earlier figure 10), or, if it was thermal water, that the steam be captured in a sauna-like "oven." Just as the new inhalation techniques required a specially designed apparatus, so too did the bottling of mineral waters. To maintain their mineral content, waters could not be exposed to the air, to light or any other impurity, nor agitated, heated, or subjected to changes in air pressure. A siphon was used to fill champagne bottles, stoppered with corks soaked in the mineral waters, and then coated with plaster. The bottling of mineral waters for drinking was thus a highly medicalized procedure that should, Lobato argued, be overseen and certified by a doctor.

At the same time that they were medicine, these early bottled mineral waters were also becoming food: they were the forerunners of the soft drinks we know today. Around 1800 a number of companies in Europe began to produce water with added CO_2, and these gained favor for their taste and their medicinal qualities, and

FIGURE 13. "Salón de embotellado de las Aguas." *El Mundo Ilustrado* 2, no. 12 (1906). With permission of Universidad Nacional Autónoma de México, Hemeroteca Nacional de México, Fondo Reservado.

were recommended for everyday use as table water. Other minerals, medicines, and drugs, such as strychnine, arsenic, quinine, and coca, were added to them by doctors, and it was also common to add sweet flavored syrup or to mix the concoction with white wine. Drinks were bottled with ever-larger amounts of sugar, part of a wider trend in the history of industrial society toward sweetened fast foods. Coca-Cola and other soft drinks based in "soda" waters (those mineral waters with carbonic acid and dissolved carbon dioxide) got their start at this intersection of medicine, fast food, and mass consumption.[5]

In Mexico, the industrial bottling of mineral waters for mass consumption began in the late nineteenth century, at the same time that the country's beer industry was founded. The sumptuous new spa at Peñón de los Baños featured a bottling plant, with the most modern and efficient machinery, using glass bottles that were produced in Monterrey's glass factories. José Lobato was a foremost proponent of bottled mineral waters for medicinal purposes, and he urged doctors and bottlers not to add red wine or other substances that would change the mineral makeup of the waters. Peñón produced plain mineral water at its plant at the bathhouse, and the Compañía Explotadora de los Manantiales del Peñón stored the bottles at its warehouse in downtown Mexico City, on Donceles Street, before shipping them out to consumers. The "Aguas Minerales del Peñón" traveled widely, earning customers across Mexico and, in competition with other mineral waters, a gold medal in the Saint Louis Exposition of 1904. It did not hurt that the medicinal qualities of Peñón's waters came recommended by Eduardo Liceaga.[6] Mineral springs, a key element of modern bathing, also figured prominently in the emergence of mass consumption of industrial products.

TEHUACÁN: STATE POWER AND THE CONSOLIDATION OF THE BOTTLING INDUSTRY

Tehuacán, Puebla, was not far behind Peñón in the popularity of its bottled waters during the Porfiriato, and after the revolution it quickly eclipsed Peñón. For centuries the town lured patients seeking a cure from kidney stones and other maladies, and since the 1890s the plush hotel and *balneario* "El Riego" received visitors by train from Mexico City and Puebla who sought an exclusive and therapeutic mineral waters treatment with baths and drinking fountains. Like Peñón de los Baños, the town's springs were promoted by the scientific community and the Secretaría de Fomento during the Porfiriato, and a tram connected El Riego to the local train station. By the early 1920s a half-dozen hotels and three bathhouses served visitors of all social strata, and Tehuacán became the foremost watering place for Mexican state officials.[7] This was due in large part to the influence of President Plutarco Elias Calles (1924–28), who made Tehuacán's Hotel-Spa El Riego a second home, occasionally even holding cabinet meetings there. Thousands of visitors synchronized their leisure choices with those of Calles and his senior officials.

The presence of the postrevolutionary government in Tehuacán soon turned into scrutiny of the healthfulness of the town's mineral waters, and in particular, its bottling industry. In the late 1920s the Departamento de Salubridad Pública made a concerted effort to regulate the production of mineral waters (*aguas minerales*), sparkling waters (*aguas gaseosas*), and soft drinks (*refrescos*). Salubridad Pública carried forward the Porfirian preoccupation with microbiology almost without pause during the revolution, and this knowledge of bacteria, amoebas, and other vectors of disease combined with much older ideas about the curative properties of particular waters to produce the conclusion that waters could hurt as well as heal. The business of bottling originally grew around mineral waters and their promise to cure, but bottled drinking water soon became desired for its purported purity, displacing in wealthy households the water delivered by urban water systems and water carriers. By the 1920s, bottled water had expanded into a thriving cottage industry producing many varieties of sweetened and carbonated waters that most often did not employ mineral waters at all. The multiplication of industrial bottled drinks resonated with deep-seated assumptions about the benefits and value of heterogeneous waters, but attracted the attention of public health officials worried about the potential harm these waters could cause.

Businessmen had been shipping Tehuacán's mineral water to Puebla and Mexico City since the late 1800s. It was initially used as a form of medicine, in line with centuries of practice of bathing in and drinking mineral waters for their curative properties, and was sold in *boticas* (pharmacies) alongside other curative waters. With the construction of the train, the water was much easier to transport, and large bottles (*garrafones*) of water were "shipped daily" for use "in all parts of Mexico, in houses, hotels, cantinas."[8] During the revolution, the largest of these bottling plants, the Cruz Roja and the San Lorenzo Mineral Water Company, were ransacked and burned, and labor mobilization troubled the industry in the 1920s.[9] In this context an array of smaller companies with improvised production methods and lax sanitary control sprouted up alongside a half-dozen bigger, more established ones. Tehuacán's bottlers continued to ship *garrafones* of drinking water to clients in Mexico City such as President Emilio Portes Gil, but doubts about quality attracted the regulatory action of Salubridad Pública.

On July 30, 1927, Salubridad Pública sent notice to Tehuacán's bottlers that they were prohibited from selling water until they could comply with Article 246 of the Sanitary Code requiring that bottled waters be free from biological contamination, and the National Railroad was ordered not to accept any water for shipment.[10] The Montt family, owners of the El Riego hotel, told the National Public Health Department (Salubridad Pública) that they bottled water from their spring only so it could be used to cure patients, but Salubridad responded that the number of *E. coli* bacteria discovered in the water bottled by them and many other companies in Tehuacán was dangerous to those patients.[11] Salubridad Pública officials made visits to the factories and found many of them to be lacking basic requirements of hygiene, with inferior capping machines and no machinery for sterilizing the water. Some bottlers ignored the ruling and continued to bottle waters and *refrescos,* but most shuttered their doors.[12] The municipal government of Tehuacán responded to the crisis in confidence toward Tehuacán's waters by arguing that "everyone in the country knows that the only thing that gives life to this city are the curative virtues of its waters," and that for years the town had been building up its credit and prestige among doctors and visitors.[13] But tourists, convinced that the waters were more harmful than curative, stayed away.

To address the problems with sanitation and hygiene, the bigger bottlers agreed to comply with national health codes by building infrastructure to capture and convey the mineral waters from the springs. After negotiating with Salubridad Pública, the companies installed a system of covered concrete canals that began at the springs and carried water to the area of town where the bottlers were. The municipal government of Tehuacán also raised funds from the state and federal governments to rebuild its own local water distribution system. By November all of the major bottlers had installed capping machines, and these were inspected and

FIGURE 14. "Empacadora—Aguilar Cacho." With permission of the Archivo Histórico de la Secretaría de Salud, Mexico City, Mexico. AHSS, FSP, SSJ, Caja 9, Exp. 1. Note the new capping machine.

certified by Salubridad Pública.[14] In addition, the bottlers agreed that Salubridad would train and certify the workers in hygienic practices, and that all products would be labeled with the date of production.[15] The bottlers coordinated efforts to raise money for the works, and installed automated production systems to reduce human contact with the water and the bottles. However, they asked in exchange that the federal government prohibit and prosecute the production of unregulated artisanal bottled water, the use of labels that falsely advertised bottles as Tehuacán mineral water, as well as the importation of foreign mineral waters.[16]

Salubridad Pública's intervention in the business of bottling extended through the early 1930s, and resulted in a consolidation of the bottling industry, with fewer artisanal producers and a narrower range of waters available to the public. Over previous decades sparkling waters (gaseosas), lemonades, and other sweetened drinks (refrescos) had proliferated, and there were, according to a spokesman for the bottling industry, "an infinite number of clandestine factories that supply the public with products of a terrible quality."[17] Often these were advertised as mineral waters but were not, and many used saccharine produced in the United States by Monsanto and imported illegally.[18] The government stepped in to control the variable quality of these waters, to the glee of large bottlers and the sugar industry. Zealous agents of Salubridad Pública sent samples to their central laboratory, then fined and closed offending factories in Mexico City, Tehuacán, Tampico, Cuautla, Aguascalientes, Tepic, Tuxtla Gutiérrez, Nuevo Laredo, and elsewhere for using saccharine, saponin, and salicylic acid in their *refrescos,* and for labeling *refrescos* as mineral waters.[19] In Tehuacán, the larger bottlers that could afford to comply with the regulations banded together to improve their infrastructure and resume production, simultaneously benefitting from the state's elimination of competitors. In Topo Chico, investment by the Coca-Cola Company provided capital for upgrading the bottling plant, allowing it to thrive in the new environment of sanitary regulation (see next section). In 1934 alone, hundreds of cases of adulteration were prosecuted by the new "Sanitary Police" of Salubridad Pública, an effort that resulted in "almost eradicating the once-frequent adulteration of drinks, especially sparkling waters and pulque."[20]

To support their effort to police the heterogeneity of waters in Mexico, Salubridad Pública mounted a propaganda campaign in the pages of the Mexico City daily *El Universal,* with articles by bottlers, lawyers, and doctors.[21] Arturo Mundet, creator of the classic apple-flavored soft drink Sidral Mundet, argued that the sugar which bottling companies put in soft drinks made them healthful, because sugar is a preservative and provides calories. That the legitimate soft drink companies used sugar was ensured by agents of Salubridad Pública monitoring their factories, and *El Universal* alerted the public that because of the cost of sugar, any *refresco* that sold for less than six cents a bottle was certain to contain saccharine or some other artificial sweetener.[22] Various articles in the *El Universal* presented cases of children intoxicated by unsanitary sweets while playing in Chapultepec Park, such as one-year-old Raúl Arriola, who died after drinking a bad artisanal soft drink. Raúl's sad story was evidence, *El Universal* argued, of the need for "strict vigilance of the streets and public spaces" by Sanitary Police.[23]

Salubridad Pública's policing of bottled waters happened at the same time that it embarked on a wider effort to ensure water quality in municipal and rural water systems across Mexico. Potable water and drainage systems were constructed with Salubridad's oversight and financing, and President Lázaro Cárdenas (1934–40) planned to almost double the portion of the national budget dedicated to Salubridad during his term, from 3 percent in 1933 to 5.5 percent in 1939. In 1935 Cárdenas authorized the Secretaría de Hacienda y Credito Publico to provide 5.5 million pesos for "supplying potable water to towns of less than twenty-five thousand inhabitants."[24] The hydraulic infrastructure promoted by Salubridad Pública delivered a singular, sanitized public water, an effort analogous to its policing of bottled waters.

TOPO CHICO: THE SCIENCE OF DISPOSSESSION

The history of the mineral springs in Topo Chico, Nuevo León, provides an example of how state hydrologists and engineers facilitated the transfer of waters from peasants to industrial capitalists. Congregación San Bernabe Topo Chico was a

FIGURE 15. Topo Chico bathhouse, c. 1890. With permission of DeGolyer Library, Southern Methodist University, Dallas, Texas. AG1987; 0643.

small town situated in a large expanse of dry ranchlands in northeastern Mexico that was awarded by the Spanish crown to Captain Lucas González Hidalgo in 1716. Residents passed the years tending livestock and farming a small cluster of fields irrigated by the waters from two local springs, a hot spring known as "Agua Caliente" and a warm spring called "Ojo Caliente." A third spring, "La Saca," issued cold freshwater but only really flowed when it rained. The springwater was scant, hot, and carried minerals, but it flowed steadily and did not hurt their fields. The water was fine for cleaning dishes and houses, and people agreed that there were therapeutic benefits from drinking and bathing in it. There was never enough water, of course, for all those who wished to use it, but while quarrels over water were common the town managed the resource in a communal fashion.

This all began to change with the building of the bathhouse in Topo Chico in 1882, and the arrival of a stream of visitors from the United States. The western United States, taken from Mexico by force in 1848, was an especially attractive destination for urbanites from the Eastern Seaboard who traveled seeking the health benefits they perceived would be gained from fresh air, wide-open spaces, and a more immediate interaction with the natural world.[25] Mineral springs, long valued for their therapeutic dimensions, were among the first places claimed, settled, and developed by the newcomers. In Texas, for example, hundreds of mineral spring resorts enjoyed a boom between 1860 and 1920, peaking in popularity around 1900.[26] The expansion of mineral and hot springs resorts across the southwest

United States was facilitated by railroad companies, which often developed the hot springs nearby their newly constructed lines.

The construction of railroads in Mexico in the 1880s enabled visitors to travel from Texas all the way to Mexico City, and Mexican mineral springs were important destinations for Americans seeking cures and seeing the sights.[27] Aguascalientes— named after the famous hot waters located there—attracted the interest of almost everyone riding the train from the northern border down to Mexico City, and traveler accounts from the time go into detail about the bathhouses and bathers of that city (see chapter 5).[28] In the border state of Chihuahua, the Atchison, Topeka, and Santa Fe line built a connection to the hot springs of Santa Rosalia,[29] and various efforts were made between 1900 and 1932 to develop the springs just south of the border in San Antonio, Chihuahua, in order to attract *gringo* tourists.[30]

The encroachment on hot springs in northern Mexico by new actors and ideas was especially notable in the small settlement of Congregación San Bernabé Topo Chico. For most of the town's history the waters of the Ojo Caliente were left to run their course, and it was not until around 1850 that townspeople constructed a six-by-twelve-meter pool out of stone and cement to store water, and a springhouse to protect the source. The pool supported the traditional domestic and agricultural uses, but also enabled a new use—bathing—especially by those arriving from afar with clear ideas about the therapeutic properties of the waters. Locals charged these health seekers a few cents for access to the reservoir, but shared no common opinion about the desirability of developing the hot springs for bathing tourism.[31] Regardless, the interest of the outsiders in the medicinal properties of the waters was keen, and their efforts to establish a business with the mineral waters were supported by Bernardo Reyes, provisional governor of Nuevo León. At Reyes's coaxing, the townspeople met and hashed out a forty-year deal by which their waters were concessioned to American Emma Slayden, who was required to build a bathhouse and provide other amenities. Four years later, A.C. Schryver of Waco, Texas, took over the contract, and hired community members to build that bathhouse using water from the Agua Caliente spring, thus creating the Compañía de Baños Topo Chico. Bernardo Reyes also awarded Schryver a concession for a mule-drawn railroad linking Topo Chico to Monterrey.

Schryver got the money to build the railroad from expatriate American financier and Monterrey resident Jules A. Randle.[32] Randle inherited the wealth of a slaveowning family from Georgia that moved to Texas just after the Mexican-American war to run a cotton plantation. He fought in the Confederate army, and upon surrender moved back to the family plantation on the Brazos River in Texas, which became one of the largest cotton farms in the region. His arrival in Monterrey in 1881 made him one of northern Mexico's largest capitalists, investing in urban railroads, properties around Monterrey, and silver mines. He was president and owner of the Monterrey and Santa Catalina Railroad and the Topo Chico Hot Springs Railroad, and owned one-quarter of the enormous Rosario Silver Mining Company.[33]

With the capital of Randle and others the business of bathing in Topo Chico was up and running. The new bathhouse had a men's area and women's area, each with its own pool 13 meters long, 5 meters wide, and 1.8 meters deep. Each of the two areas also had twelve tubs, fabricated of wood and zinc, each in its own 3-by-2.5-meter wooden stall with a wood and cloth cot. Admission to the bath, including the 45-minute, three-mile tram ride from downtown Monterrey, was 50 cents. Nearby, another group of Americans built a luxurious hotel to cater to the American tourists, with a kitchen run by an American chef. In 1893 E.R. Glass built the Hotel Marmól across the street from the bathhouse to cater to the new influx of visitors to the hot springs, by then known regionally, nationally, and internationally for their curative properties.[34] Jules Randle invested some of his silver fortune in the $250,000 Hotel Marmól.[35]

A number of German and American doctors arrived to Topo Chico to offer their services to health-seekers using the waters.[36] One of these, Dr. G.F. Brooks, moved from the paradigmatic mineral springs town of Hot Springs, Arkansas, to try his luck at this emerging tourist health spa.[37] So widely known were the springs that J.H. Blackburn, a doctor from Texas searching for a cure for his gout and diabetes, included Topo Chico in an itinerary that also listed far-flung mineral water health resorts such as Lithia Springs, Virginia, and Hot Springs, Arkansas.[38] The Mexican National Railroad Company, which connected the United States to Monterrey and the local Topo Chico tramway, distributed a free booklet promoting "Tropical Tours to Toltec Towns," and highlighted Topo Chico's "superb baths and a good hotel, all under American management."[39] The town of Topo Chico quickly became a mecca for American visitors, giving rise to a host of peripheral services, such as a local dairy run by American settlers. One visitor noted that "the whole settlement" of Topo Chico was "managed by Americans."[40]

Bottling was an equally important business at Topo Chico that grew to eventually displace the bathhouse in the 1930s. The waters of Topo Chico achieved such fame during the last decades of the nineteenth century that Randle contracted the rights to six liters per second of the springflow from the community of Topo Chico and began bottling the mineral water under the brand name of Topo Chico for distribution to visitors and inhabitants of the region.[41] In 1900 the community of Topo Chico signed a contract giving permission to Emma Slayden to build a bottling plant, although still in 1902 a traveler noted that "the springs themselves stand in a shady grove" and were not captured by a bottling plant at their origin.[42] Emilio Hellión, a Frenchman residing in Monterrey, bought into the Topo Chico bottling company and, together with Manuel Cantú Treviño, secured capital from the New York firm Wilson and Company to expand and consolidate the operation.[43] At the same time, Pedro Treviño, one of San Bernabé Topo Chico's wealthy landowners and owner of the ephemeral La Saca spring, built a spring house and factory for ice and soda, investing upward of 100,000 pesos. Much of this money likely came from outside investors.

Conflicts emerged as bottling intensified. Treviño's development of the La Saca spring was opposed by members of the community, and as a result of their complaints an expert in hydrology was sent by the city government of Monterrey to investigate. Because he had dug a well and the springwater did not flow beyond his property, Treviño was found to be the legal owner of the spring. This first conflict over the springs set the tone for the competing social uses and politics of these springs during the next forty years, which would continue to be characterized by divisions within the community and a major role for government scientists in determining the nature of the water resources, and in transferring control and use to capital.[44]

As bathing and bottling grew in popularity at the end of the nineteenth century information was needed to govern competition over mineral springs. The emergent science of hydrology assumed the task of determining if water was property of the nation, the state of Nuevo León, or private landowners. During the rule of Porfirio Díaz (1880–1911) the government made an effort to map the Mexican countryside, and distributed lands to surveying companies to promote this activity.[45] Despite these actions, small water sources such as hot springs remained under the radar of the state. For example, in 1904 local residents asked the secretary of agriculture and development for rights to build a bathhouse at the hot springs in Las Cabras, Chihuahua, but the federal government could not even find those hot springs on their map.[46] In the case of the hot springs near Catemaco, Veracruz, the Secretaría de Obras Públicas did not possess a map of the region, let alone of the springs, and could not acquire one from any other branch of government.[47] Even when the government's own maps registered hot springs, and those springs were located on federal lands, officials usually had no information about spring flow, temperature, established uses, or anything else.

The waters of Topo Chico attracted the attention of regional and international capital, and their status as a community resource was challenged. In 1898 the national government declared the waters of the drainage where the Topo Chico springs were located—the Arroyo Topo Chico—to be national waters, because they led to the Santa Catarina River, which eventually emptied into the Río Bravo.[48] This was confirmed, at least on paper, by a map from 1904, although the community of Topo Chico continued to dispose of "its" hot springs in the ways established during the previous centuries: for domestic use and gardens, for animals, and for bathing. The recent turn to concessioning the waters to bottling and bath companies did not put into question community ownership over the resource.[49] During the revolution, popular ideas about "land and liberty" reinforced local control of the hot springs and in 1917, a new Constitution was written which enshrined the radical liberal idea that the land should belong to those who worked it. In 1918, with local agrarian rebels in charge of the bathhouse, the waters of the drainage in which the Topo Chico hot springs were located were ruled to be private rather than national waters. This ruling validated the existing contract between the Community of San Bernabé Topo Chico and the bottling and bath companies, and short-circuited the possibility that the waters would be national-ized by the federal government.[50]

Nature in northern Mexico did not submit readily to the scientists, for the arid landscape did not conform to hydrological concepts such as "river." Water often only flowed during the rainy season, and small drainages (*arroyos*) such as that of Topo Chico would only carry water during storms. The same maps that failed to register hot springs depicted flowing rivers that were in reality simply drain-ages that hardly ever carried surface water. Furthermore, water laws written before the rise of hydrological science did not contemplate the connections between the surface waters and subsoil waters,[51] and Mexico's constitution only incorporated groundwater in 1945, with a reform to Article 27.[52] To complicate this issue, the waters of hot springs, which emerge from deep below the surface of the earth, usually have little to do with those that run in drainages either as subsoil water or surface water.

Water was considered a common-pool resource in Mexico before and after the revolution, as the postrevolutionary state incorporated popular concepts of com-mon property of land and water into the new Constitution of 1917. But who had the authority to designate the legitimate users of that common property? The answer involves issues of scale and scientific authority. Mexican water administration was organized legally by a principle of geographical scale. Water that did not flow beyond the boundaries of a single property was considered part of that property. Water that flowed across different properties but not across a state's borders was under the jurisdiction of that state's government. That which crossed state lines, such as the water carried by the Salado and San Juan rivers and their tributaries, was national; if a river drained into the Río Bravo (known as the Río Grande in the United States) it was water governed by international treaties as well. All national water was the common property of the nation, to be administered by the federal government, and during the revolutionary and postrevolutionary period, water, like land, was the object of nationalization and redistribution by the federal gov-ernment. These were scales of government, and obviously political.

Science supported the slow process of primitive accumulation and the transi-tion from peasant uses of water to capitalist uses. In Topo Chico, a local spring that in 1880 supported diverse economic activities of peasant households was, by 1950, completely utilized by one the biggest industrial bottling companies in Mexico and the world. Rather than hinder it, the long process of revolution (1910–20) and postrevolutionary state formation ushered along the process of accumulation by dispossession. The armies and leaders of this conflict formed constantly shifting alliances, and communities were divided along these lines. In Topo Chico the rev-olution fractured existing agreements about the legitimate uses and owners of the spring waters, and a group of rebels rose in opposition to those in the community who dominated the land and water and controlled the town government. As the

FIGURE 16. "Manantial Agua Caliente," c. 1930. With permission of the Archivo Histórico del Agua, Mexico City, Mexico. AHA, AN, Caja 463, Exp. 4893.

revolutionary movement across northern Mexico died down, and the victorious generals began the process of rebuilding the Mexican state, the local rebels of Topo Chico adopted the politics of agrarian reform (*agrarismo*), pressing the federal government to nationalize land and water held by the wealthier members of the community and award it to them as a collective farm, or *ejido*.

The social upheaval wrought important changes to the bathing and bottling businesses that used Topo Chico springwater. Pedro Treviño's ice and soda factory, which utilized the La Saca spring, was abandoned, and the foreign investors in the Topo Chico bottling company fled, selling their stakes to regional businessmen Manuel Barragán and Leónides Páez. The Compañía de Baños met the same fate when the national and international tourism that had supported the bathhouse and hotel ceased completely because of the violence. In 1921, in one of its first actions, the newly constituted Department of Public Health (Salubridad Pública) closed the baths, citing the unhygienic state of the facilities.

In 1922, the contract between the town of San Bernabe Topo Chico (still the holder of legal rights to the hot springs water) and the Compañía de Baños expired.[53] Without a contract for the waters, without a bathhouse in condition to receive customers, and without customers brave enough to visit Topo Chico, the Compañía de Baños went out of business and the installations were taken over by *agraristas*. They, however, had no means with which to improve or maintain the infrastructure of the baths, and soon "the roofs were falling and the tubs, walls and pipes were so deteriorated and filthy that very few people dared use them."[54] In the turmoil, the town government asserted itself, taking over the administration of the hot springs water "by the unanimous will of the neighbors and community members who live in Topo Chico."[55] In an effort to force the bottling company to agree to a new contract, the town government cut off water to the bottling plant and took out advertisements in the newspapers of Monterrey accusing the company of bottling regular water, not mineral water.[56] Soon after, the town government delivered a petition to the federal government in which it claimed to be the rightful owner of the mineral springwater and asked that it be returned. In 1924 the town reopened the baths under its own control after correcting the problems cited by Salubridad Pública.[57]

The struggle over land and water in Topo Chico proceeded in fits and starts, and different levels of government intervened on behalf of different actors. To deal with the *agrarista* uprising, in December of 1923 the governor of the state of Nuevo León orchestrated a land transfer outside of the federal agrarian reform process aimed at establishing peace between the competing factions in the town. A transfer shifted 1,444 hectares of land acquired by large landowners in the mid-nineteenth century to the *agraristas*, but before this agreement was signed into state law in March 1925, the community submitted a parallel request to the federal government's Agrarian Reform Commission (CNA) for the return of those same lands, claiming that the Congregación San Bernabé once owned them. The local branch of the federal government approved the request, but it was rejected at the state level by the governor of Nuevo León, who had already brokered a similar reform. Pressured by the federal government, the state government eventually approved the federal creation of an *ejido* as a new concession of land rather than a return of land. In August of 1926 President Plutarco Elias Calles declared a resolution awarding the *ejido,* and thereby annulling the state of Nuevo León's 1923 agreement.[58] This award of land rejected the community's ancestral claim to the resource, and reinforced the federal government's position that it was the only legitimate owner and administrator of national land and water.

Once the land was delivered, the struggle turned to water, and was fought on the terrain of hydrology. The central problem was that there was not enough water to irrigate the newly distributed lands. The Presidential Resolution of 1926 parceled out 25 hectares of gardens and orchards near the town, and 2 liters per second of water from the hot springs for domestic uses and for livestock, but did not provide the 7.9 liters per second of water needed to irrigate those 25 hectares. A bigger problem, however was that the resolution also failed to provide the 73.2 liters per second of water needed to irrigate the 1,444 hectares of previously unirrigated lands that was also part of the distribution.[59] With the hope of resolving this problem, the community of Topo Chico petitioned the secretary of agriculture to declare the waters of the Arroyo Topo Chico national, and not private, so that they might lodge a claim to them through the federal government's agrarian reform process.[60] The secretary of agriculture sent an engineer to make a study (the second) of the springs and the Arroyo Topo Chico into which they drained, and in

June of 1927 the waters of the arroyo, including the springwaters, were indeed declared national property because, the engineer argued, the waters formed part of a drainage that eventually led to the Río Bravo.[61] Once placed under control of the federal government, the issue turned to whom the federal government would award their use.

When the hot springs waters were nationalized (for the second time), the local town government of San Bernabe Topo Chico immediately took over the bathhouse. Its leader, Celso Cepeda, asked permission from the federal government to "make use of the hot water for the public baths that [the town] will refurbish using money from the agrarian bank."[62] The town government then squared off against the Compañía de Baños Topo Chico, accusing it of never paying the monthly charge for the waters of 100 pesos that was stipulated in the contract. The Compañía countered with the opposite claim: that it had been paying the 100 pesos to Cepeda for some time.[63] Then, in March of 1928, the *ejidatarios* of Topo Chico occupied the bathhouse.[64] The state government of Nuevo León immediately intervened, ordering the Congregación to return the facilities to J.T. Garza, proprietor of the Compañía de Baños.[65] The state government declared that the *ejidatarios* did not have permission to use the waters for industrial purposes, and the Compañía de Baños could therefore continue to use them for bathing and bottling.[66] This decision was based on the assertion that the hot springs were local waters rather than federal waters.[67] The federal government protested to the state that "the declaration of Arroyo Topo Chico as national waters would not be reconsidered."[68]

The state government continued to assert its right to manage both the Topo Chico springs and the conflicts surrounding them, brokering a deal between the town of Topo Chico and the Compañía de Baños de Topo Chico and its operator, J.T. Garza. In a twenty-year contract signed in May of 1928, the town was declared owner of the bathhouse, with its baths and offices, as well as a nearby park and bandshell and various other properties. These facilities were to be rented by the Compañía de Baños Topo Chico for 100 pesos a month. The waters of the hot springs were to be used only for the bathhouse, and then sent to a tank where the town could distribute them for irrigation. Garza was obliged to invest 20,000 pesos in repairs over the next five years.[69] The town made a separate, forty-year (1928–68) contract with Manuel Barragán for the use of the waters by the bottling company—the Compañía de Aguas Gaseosas.[70] The *ejidatarios* of the Topo Chico were told to relinquish their hold on the bathhouse and spring waters, and that there was no water in the Río Santa Catarina to irrigate their new fields.[71] The most they got was permission from the secretary of agriculture and development to build, at their own cost, a horizontal filtration well (*galeria filtrante*) to collect the water.[72] They made an effort to secure an industrial concession for the hot springs water, presenting a map from 1904 that showed the hot springs were part of the Río Santa Catarina, and thus national waters they could solicit.[73] Bathhouse operator J.T. Garza defended his access to the water with a municipal map of Monterrey that showed the Arroyo Topo Chico petering out in the irrigated fields of San Nicolás, without reaching the Río Santa Catarina. It was not federal water, he concluded, and therefore ownership by the town, and the lease to the bathhouse and bottling companies, should stand.[74]

For most of the 1920s both Nuevo León and the federal government of Mexico claimed jurisdiction over the springs, using scientific arguments about the origin and destination of the waters. The contracts brokered by the state government of Nuevo León were based on rights and concessions that had yet to be established by the federal government, which by then considered itself the proprietor of the water. In order to award these concessions, and regularize the contracted uses of the water, in August 1929, the Federal Secretaría de Agricultura y Fomento sent engineer Ramón Áviles to conduct a third study of the springs. He spoke with different parties that used the hot springs, took measurements of streamflow and photographs of the installations, drew up maps of the site, and wrote a detailed report. He concluded that both the Ojo Caliente and the Los Baños (Agua Caliente) hot springs were permanent, and the La Saca flowed only when it rained. The Los Baños (Agua Caliente) hot spring was used by the bathhouse, the bottling company, and the townspeople for domestic chores, while Ojo Caliente and La Saca were used to irrigate gardens and orchards. All the water from the three sources was used completely.[75] Áviles's report concluded that the Topo Chico springs were national waters, and the 1926 presidential declaration of water rights should stand. The town had rights by presidential decree to 2 liters per second (lps) of the Los Baños (Agua Caliente) hot spring for domestic uses. In addition, the engineer assigned 7.92 lps of the water divided among La Saca, Ojo Caliente, and Los Baños to irrigate the twenty-five hectares of orchards and fields for which there was previously no water assigned.

With the submission of Áviles's report, any water use that was not formally recognized by the federal government's secretary of agriculture became illegal, including customary uses that had been practiced by townspeople for generations. Furthermore, with nationalization of the water confirmed by the report, whatever water not assigned by the federal government was up for grabs through a process of concession. Mexican water law held that rights to nationalized water should be awarded to those who had established continuous, peaceful use of that water during the previous five years. According to this formulation, both the town of San Bernabe Topo Chico and the companies could lay claim to the liquid: the water passed through the bottling plant and baths, and then the community used it. Except for the water that ended up inside the bottles, the bathhouse and bottling plant made "nonconsumptive" use of the liquid and handed it over to the community for domestic uses and agriculture.

The nationalization of the Topo Chico springs directly benefitted the companies, and facilitated the long-term shift in control from peasants to industrial capitalists. Shortly after the 1929 report was submitted, the Ministry of Agriculture and Development alerted the bottling and bath companies that they would need to solicit a water concession or confirmation of existing use or their access to the springwater would be suspended.[76] In the same month that the engineer made his survey, the Compañía Topo Chico filed a request that the government recognize its rights to the springwater, claiming that it had used the medicinal waters in the bathhouse since 1886.[77] For its part, the town of San Bernabe Topo Chico filed a request for a new concession of waters, arguing that it wished to expand the bathhouse to expand curative services to a "public in pain."[78] At that moment, however, the Ministry of Agriculture and Development overrode the deal brokered by the state of Nuevo León that gave the town the property rights to the bathhouse. The federal government ruled that the owner of the bathhouse was Garza, not the town, and that furthermore he had "acquired the rights to the use of those waters."[79] Also, a concession of 1.396 lps of water from all three springs was awarded to the bottling company, and it was advised that it should no longer pay the 100 pesos a month to the town for the use of the water, for the town was no longer the owner.[80] The town, seeing the water of the hot springs slip from its hands, demanded its return, accusing the governor of Nuevo León of arbitrarily given water away to "outsiders."[81]

The consolidation of capital's control over the Topo Chico springs in the form of bathing and bottling moved steadily forward despite, and even because of, the revolutionary turmoil and political uncertainties of the teens and early twenties. The reconstruction and strengthening of the nation-state in Mexico carried with it the nationalization of property rights for land and water and, in cases such as Topo Chico, the state facilitated the transfer of common resources to private firms. The Topo Chico bottling company actually expanded its offerings during the revolutionary years to include flavored sodas such as ginger ale ("Yinyereil") and an apple drink called "Eva." It also improved its factory by investing in a metal bottle capping machine.[82] And, in 1926, the company became the first bottler in Mexico to produce Coca-Cola.[83]

By 1930, after years of neglect, the Compañía de Baños had rehabilitated the bathhouse by laying down tiles and providing mattresses and rugs, and had spruced up the town park, which had been used by the *agrarista* rebels to graze their horses. Once fixed, a stream of visitors—including foreigners—returned to the baths, lured by their medicinal qualities.[84] Salubridad Pública monitored the installations, to assure cleanliness and attractiveness for the tourists to the springs, and told the community to scrub the tank where the residual waters from the bottling and bathhouse collected before being sent to the fields.[85] Bitter residents replied that the only reason it was dirty was because the bottling plant dumped syrups, soap, label glue, and machine oil into it, and demanded that their water be delivered to them first and to the bottling plant later.[86] Having already won the day, the engineers of the Ministry of Agriculture responded with righteous indignation, labeling the complaints "morally wrong" and calling the townspeople "liars."[87] The *ejidatarios*, for their part, concentrated their energy on fighting with agricultural producers from neighboring communities for the water of the Río Santa Caterina.[88]

During the next decade, access to the Topo Chico hot springs would narrow even further, as the bathhouse closed due to fading public interest and the bottling industry consolidated its hold over the water. In 1930 the bottling plant of the Compañía Topo Chico entered another period of expansion, and began to export products by road and rail to cities in the states of Nuevo León, Tamaulipas, and Coahuila. The company substituted the older brand of ginger ale with a new product called "Ginger Ale Topo Club."[89] The Coca-Cola Company strengthened its relationship with the Compañía Embotelladora Topo Chico, and its products, introduced in 1926, led the growth. Attracted by the success of the Topo Chico Company, a competing bottling firm pressured the federal government to reassess spring flows once again (the fourth time), and then grabbed the newly identified unassigned water before it arrived to the townspeople, who were now the last in line. Local control of the water for agriculture, drinking, and bathing was no more. Like most cities in Mexico, Monterrey grew rapidly after 1940, incorporating neighboring communities and their lands, and the community of San Bernabé Topo Chico was integrated into the urban sprawl in the 1960s.

CONCLUSIONS

Governments in Mexico facilitated the business of bottling in Mexico's mineral springs after the revolution (1910–20). In Tehuacán, the federal government's Departamento de Salubridad Pública imposed rules of sanitation and hygiene that helped consolidate the control of large businesses over the production of mineral waters, *refrescos, aguas gaseosas*, and other diverse bottled drinks. This resulted in the control of the bottling business in that town by the Garci-Crespo company. In Topo Chico, the government deployed the science of hydrology to gradually wrest the spring waters from peasants and ranchers and redirect them to a bathhouse and bottling plants. The Coca-Cola Company eventually took over bottling at the Topo Chico springs, propelling the eventual displacement of all other social actors and uses.

In both Tehuacán and Topo Chico the end result of this dispossession can be seen in the meanings that these two place names carry today: both these words are brand names synonymous for bottled gasified waters. By the 1950s, state intervention in these two mineral springs helped transfer the waters from the hands of local peasants to large industrial companies. Tehuacán's Garci-Crespo company, founded in 1928, grew to be the most important producer of mineral waters in central Mexico, becoming the Peñafiel company in 1948, and eventually forming part of Cadbury Schweppes (1992), and then part of the Dr. Pepper Snapple Group (1995). The spring still bubbles forth deep beneath the bottling plant in a carefully controlled catacomb that can be visited by tourists (but not photographed). In northern

Mexico a sparkling water is a Topo Chico, a name that is increasingly used in Texas and the rest of the southwestern United States as well. Topo Chico is now the founding brand of the Arca Continental Company, a conglomerate that produces snack foods and is the second largest bottler of Coca-Cola in Latin America.[90]

Hydrology and biology accentuated the homogenization of waters, but paradoxically, they also made their heterogeneous qualities more attractive. In Tehuacán, the dangers of biological contamination were identified and combatted, reducing the variable quality of bottled waters, contributing to their standardization and turning them into a kind of public water. At the same time, the mineral waters of Tehuacán and Topo Chico in their commodity form of bottled drinks enjoyed widespread appeal precisely because of their specificity: their mineral content, their particular geological origins (real or imagined), and the idea that these springwaters have unique, culturally rich histories.[91] The commoditization of these waters did not smooth over their specificity; in fact, it depended on it.

Spa Tourism in Twentieth-Century Mexico

Mineral springs had been a centerpiece of the experience of foreign visitors to Mexico since the colonial period, when Spanish clerics soaked in the waters of Peñón de los Baños and built baths at Cuincho and San Bartolomé, and natural scientists sought to distill their curative secrets in the field and the laboratory. With the demise of the Spanish empire, travelers made their way more freely to Mexico, often stopping at different mineral and hot springs to test the waters. The war with the United States in the 1840s brought a few soldiers from the United States to explore the country. After 1860, foreigners rode the rails into Mexico as tourists, searching for novel experiences of food, music, people, crafts, the splendor and bustle of Mexico City, and archaeological sites such as Teotihuacán.[1] The number of travelers' accounts exploded after 1880, testimony to the new transportation infrastructure, as well as the formation of a leisure class in the United States and Europe. This expansion of tourism included wealthy Mexicans, a tide that also lifted workers and even rural dwellers as the twentieth century progressed.[2]

Waters were a principal tourist attraction, and the economy of leisure was built around them.[3] Bourgeois residents of European and North American cities had been spending their vacations at mineral water spas and seaside resorts for most of the nineteenth century, and with the opening of Mexico to travel these same groups began to visit springs in Aguascalientes, Tehuacán, Topo Chico, Lake Chapala, and other sites. In chapter 5 we saw that as early as the 1840s urban Mexicans took to country baths on hot spring days, and the bathhouses of Mexico's towns and cities were famous in the second half of the century for their number and quality. Elite Mexican bathers joined foreigners at watering places that were oriented toward a wealthy, cosmopolitan clientele, and rustic mineral springs bathing establishments served those with less money who were also caught up in the spa boom. As was the case with bottling businesses, spa tourism also generated conflicts over property and access to mineral waters.

In this chapter I discuss the business of mineral springs bathing between 1920 and 1960. While springs such as Topo Chico and Tehuacán were captured by industrial capital for bottling, many others were developed as tourist bathing destinations. As we have seen, the business model of developing mineral springs for health and therapy was pioneered in the Valley of Mexico by investors close to President Porfirio Díaz, including his doctor, Eduardo Liceaga, and his father in law, Manuel Romero Rubio, who built bathhouses and bottling plants that took advantage of the special properties of the waters of Guadalupe and Peñón de los Baños. With the consolidation of the postrevolutionary state in the 1920s in the hands of northern Mexican generals, the effort to develop Mexico's heterogeneous waters was carried out with even more urgency. Not only did these politicians promote this development with new water laws and government resources, they also invested their own money. The dispossession of mineral springs brought about by government officials benefitted the businesses of those very same people.

Another notable aspect of this period in the history of Mexico's waters is a shift in the rationale of mineral springs bathing away from health and therapy and toward leisure and tourism. By the 1920s the fascination with mineral water cures was waning among doctors, and the postrevolutionary state was more interested in promoting public health through hygiene and sanitary water infrastructure. The microbial revolution of the nineteenth century propelled biological understandings of disease and wellness to the fore, a position that was consolidated after World War II with the development of antibiotics. While most people retained the idea that mineral spring waters were curative, they came to view waters first and foremost as vacation destinations that contributed to overall well-being through rest, relaxation, and exercise. The therapeutic efficacy of the waters became less important.

Mexico's water history is full of conflict generated by primitive accumulation, policing, and scientific debate. However, the literature on hot springs, almost all of it focused on Europe, scarcely mentions questions of access and property, or struggles among peasants, scientists, clergy, and capital for control of waters. Eric Jennings (2006), however, suggests that these struggles were a central issue for mineral springs. In Mesoamerica, lands and waters were occupied by peasants long before the arrival of Europeans, and although these peasant communities coexisted with and contributed to states for more than a thousand years, they retained some measure of autonomy over local resources. Like that of Topo Chico, the history of the town of Ixtapan de la Sal, in the state of Mexico, provides ample evidence that encroachment by capital and the state on these peasant waters was met with resistance. Unlike Topo Chico, however, in Ixtapan de la Sal that encroachment was eventually limited by this resistance. There, the municipal government of the community fought the alienation of its waters, producing in the end a compromise in which capital developed some springs for an elite market, and the municipal government controlled other springs, ensuring that locals and other less wealthy visitors had access to them, and that the town itself benefitted from the business of bathing.

THE *SONORENSES* AND MINERAL SPRINGS TOURISM

Agua Caliente hot spring bubbles up on the southern bank of the Tijuana River just a hop, skip, and jump from the United States, across a seasonal trickle of water. The spring is located on what was once the sprawling Rancho de la Tia Juana, a property acquired by Santiago Argüello in the 1840s just before the present-day border was established by the Treaty of Guadalupe Hidalgo. The waters of Agua Caliente were used by locals as medicine, including to ease childbirth. One testimony in 1920 claimed that all "the recent generations and descendants of Don Ignacio Argüello saw the light for the first time in those waters."[4] Inspired by the success of hot spring resorts across the American west in the last decades of the nineteenth century, many speculators had designs on the springs, and in 1899 the Argüello family leased the springs to David Hoffman, who founded the "Agua Caliente Sulphur Company" with a bathhouse and hotel that became well known among tourists to California. Revolutionary conflicts beginning in 1911 kept visitors at home and a major flood in 1916 swept away the buildings. Interest in the property remained, but the Argüellos retained ownership of both the springs and a makeshift changing room located nearby, charging 25 U.S. cents to anyone who wished to use them. In February 1921 a Tijuanense named Rodríguez Galeana, allegedly "conniving" with Americans, filed a claim with the federal government for the springs, arguing that they were national waters because they flowed within the banks of the Tijuana River, and providing as evidence photographs of the 1916 flood that filled the river and covered the springs.[5] The Rodríguez Galeana water grab was rebuffed, but the boom in the business of mineral springs bathing and the 1920 prohibition of alcohol in the United States made the Agua Caliente springs too attractive to leave undeveloped for much longer.

In 1926 three wealthy Americans teamed with Baja California governor Abelardo L. Rodríguez (1923–30) to purchase the springs and a large parcel of land around them from the Argüellos and build a sumptuous spa, hotel, restaurant, and casino complex, which opened its doors in June of 1928. Soon in Agua Caliente there was a swimming pool, private bungalows, a horse racing track, a golf course, and an airport to receive American tourists. Agua Caliente's alcohol, gaming, and nightlife attracted Hollywood's famous and wealthy, as well as political figures and gangland notables such as Lucky Luciano and Al Capone. As governor, Abelardo Rodríguez ensured the success of Agua Caliente by smoothing political wrinkles, facilitating permits, and negotiating with labor organizations. He also invested his own money in the enterprise, and dedicated state resources

FIGURE 17. "Manantial del Agua Caliente sobre la Margen Izquierda y dentro del Cauce del Río de Tijuana," c. 1920. With permission of the Archivo Histórico del Agua, Mexico City, Mexico. AHA, AS, Caja 723, Exp. 10513.

to the construction of roads, the provision of electricity, policing, and other infrastructures and services.

Rodríguez was a fellow *sonorense,* revolutionary, and close associate of General Plutarco Elias Calles, who would be president from 1924 to 1928. Rodríguez rose from poverty to become military commander of Baja California in 1921 and also governor of that federal territory from 1923 to 1930. After Calles stepped down from the presidency in 1928, he named a series of his allies to the position, including Rodríguez. Rodríguez brought his experience with hot springs tourism to Mexico City when he was named to take over the presidency after Pascual Ortiz Rubio resigned halfway through his term. Rodríguez was installed as interim president between 1932 and 1934 precisely because of his loyalty to Calles and his followers, and quickly became involved in promoting the ex-president's favorite water tourism spot: Tehuacán, Puebla.

Calles and the political class of Mexico City popularized Tehuacán in the 1920s, and by 1930 the city was bustling with tourists. Bathing in mineral springs and feasting on the regional delicacy *mole de cadera* became symbolic of the privileges of the new revolutionary elite that Mexican tourists hoped to emulate. With so much money pouring into Tehuacán, it also became known for drinking, gambling, and prostitution, activities that Rodríguez viewed as a normal part of the tourist business.[6] President Rodríguez further promoted the hot springs resort town by dedicating federal money to building a paved highway from Puebla that reduced the time in transit from Mexico City to a mere four hours. It was rumored that both Rodríguez and Calles would stand to personally benefit from the highway, for they had invested their own money in the mineral springs bottling plant built by José María Garci-Crespo and in the luxurious spa hotel that he opened for business in 1934.[7] This hotel—the Garci-Crespo—eclipsed the recently built Hotel Casino de la Selva in Cuernavaca as the most luxurious watering spot in the Americas.

When Lázaro Cárdenas assumed the presidency in 1934 he stepped back from the tourism development model, devoting state resources instead to productive activities, especially agriculture. Aiming to marginalize Calles and the *callistas*, and to limit the negative dimensions of tourism development, he outlawed gambling, which was a principal economic interest of that group. Cárdenas closed the Agua Caliente casino complex in 1935, but he actually supported a reformed version of tourism development in the rest of the country.[8] The Hotel Garci-Crespo thrived without gambling, and, rebaptized as the Hotel Peñafiel in 1948 (a name shared by the brand of mineral waters bottled on the site), continued as the premier vacation spot in Mexico until the development of Acapulco by President Miguel Alemán, who, expanding the *sonorense* water tourism development model, plowed federal resources into building up that coastal resort for international and national visitors in the late 1940s and 1950s.[9]

In one way or another, all the postrevolutionary presidents and politicians viewed tourism as a desirable development strategy, and many combined their state roles with personal investments. Pascual Ortiz Rubio, who served as secretary of communications and public works in 1920–21, and as president for the two years preceding Abelardo Rodríguez (1930–32), formed the Compañía Impulsora de Acapulco with the goal of building a tourist hotel in that town. In 1935 Ortiz Rubio created Campos Mexicanos de Turismo (CMT), which operated a hotel for U.S. tourists on the Pan-American Highway in Ciudad Valles, San Luis Potosí. A few years later, CMT bought a hot springs bathhouse and hotel in Ixtapan de la Sal, with the plan to turn it into Mexico's premier mineral springs resort and an international tourist destination.

PARTING THE WATERS IN IXTAPAN DE LA SAL: COMMUNITY, CAPITAL, AND THE STATE

Where politicians and investors eyed profits in mineral springs such as Agua Caliente, Tehuacán, and Ixtapan, locals saw waters that had been used by their families and communities longer than anyone could remember. These were very different understandings of the value of waters, and in the town of Ixtapan de la Sal that difference led to ongoing resistance to the development plans of businessmen and politicians. Ixtapan de la Sal is today a tourist town lodged in the hills that descend south from the Nevado de Toluca volcano in the state of Mexico, about two hours away from Mexico City on the highway. It has scores of hotels that serve a wide range of visitors, from wealthy residents of the cities of Toluca, Cuernavaca, and Mexico City to humble *campesinos* from nearby towns. The landscape surrounding Ixtapan is dotted with country houses for those who can afford them, and growing neighborhoods for those who work in the town's thriving tourist economy. People come to Ixtapan de la Sal to enjoy the sun, the climate and most importantly the waters. There are a multitude of swimming pools in hotels and private houses, the Ixtapan Aquatic Park with its hot springs bathhouse, as well as the hot spring pools of the Municipal Bathhouse in the town center.

None of this existed before 1930. For most of their history, the waters of Ixtapan de la Sal's multiple hot mineral springs were used for producing salt, and although local people almost certainly bathed in the waters, there is no evidence that anyone else did. There were five springs of importance, as well as a handful of tiny ones. San Gaspar, El Bañito, and Santa Catarina springs were located near the town, and the first two were developed early on into rustic pools. Four hundred and fifty meters uphill to the north of town, a pond called Laguna Verde, also fed by hot springs, was used by those bathers who sought a cure for communicable diseases. The El Salitre spring was located half a kilometer south of Ixtapan, downhill toward the town of Tonatico (which has its own springs). Freshwater was taken from a small source near the San Gaspar springs, and, beginning around 1878, an eighty-kilometer-long canal brought freshwater from the slopes of the Nevada de Toluca to the towns in the region.[10]

The saline waters of the mineral springs of Ixtapan left thick deposits of salty soil, and these, as well as the waters themselves, were used to make very pure, white table salt that was sent as tribute to the Aztec kings by the Matlatzincas, and in the colonial period was used for refining silver from the mines of Taxco and Zacualpan to the south. The town had a church, and beginning in 1822 was the seat of municipal government, but had very little freshwater, agriculture, and inhabitants. Salt-making using evaporation ponds—called "Ixtamiles" ("salt-fields" in Nahuatl)—was still an important activity at the end of the nineteenth century, but by 1930 was no longer a profitable business, and only lived on in artisanal fashion to supply local markets based in barter.[11] While the salt was a key product of the town of Ixtapan, archival records indicate that the conflicts that arose were focused on the use of land for the production of food and livestock.[12]

Between 1870 and 1930 the salty waters of Ixtapan slowly strayed from the orbit of silver mines and were integrated into the tourism economy as the bathing boom that swept Mexico and the world accentuated the value of Ixtapan's heterogeneous waters. In 1877 the Italian immigrant José Nosari identified Ixtapan de la Sal as a good place for a spa that would ride the wave of popularity in mineral springs bathing. He forged a contract with the municipal government for the rights to use the mineral springs for 99 years, paying 100 pesos annually. Nosari sold this contract in 1890 to Santiago Graf, a Swiss immigrant with similar pretensions who managed the hot springs bathing business in Ixtapan de la Sal until revolutionary soldiers

FIGURE 18. "El Bañito: Al Fondo el Campo de Aviación, Ixtapan de la Sal, Mex." Courtesy of Luis René Arizmendi, Ixtapan de la Sal, México. Date unknown. Note the saltworks in the background.

FIGURE 19. "El Bañito: Una Pileta Circular." With permission of the Archivo Histórico del Agua, Mexico City, Mexico. AHA, AS, Caja 2058. Exp. 31075, pp. 187–96 (June 2, 1941), "Informe #79."

attacked the town, killed five townspeople (the "martyrs of 1912"), and drove the tourists away. Graf abandoned his business between 1912 and 1918 and could not pay the accumulated rent, so the municipal government contracted with a series of three Mexican businessmen between 1919 and 1930 to exploit the hot springs.[13]

José Reynoso, an engineer working in the silver mines of Zacualpan in the 1920s, was impressed by the potential for a bathing business in Ixtapan and had the governor of Mexico State, Coronel Filiberto Gómez, broker a deal with the municipal president José Vergara to acquire the rights to the hot springs, as well as the right to build hotels and bathhouses on municipal land and to use freshwater for these tourism projects. In turn, Reynoso promised to deliver fifteen percent of his profits to the municipality.[14] The state government spent 4,500 pesos to renovate the pools at the San Gaspar springs, and Reynoso pitched in 1,500 pesos for dressing rooms. Reynoso dedicated much more of his money to buying land and building a 40,000-peso, 34-room hotel next to the San Gaspar springs, which came to be known as the Hotel-Baths of Ixtapan de la Sal.[15] State and municipal governments also contributed to developing a much smaller pool at the El Bañito spring that was simply a circular, four-meter-wide hole dug in the earth. The remaining undeveloped springs were also used for drinking and bathing by visitors attracted to Ixtapan by the curative properties of the waters. Reynoso built a rectangular pool at the Santa Catarina spring, with dressing rooms, for the locals.[16]

Reynoso's Hotel-Baths of Ixtapan de la Sal brought an unprecedented influx of visitors to the town, despite the fact that the road was still a dirt track, and his business flourished, serving as many as a hundred people a day by 1941. Other hotels and *pensiones* for tourists sprouted up during the 1930s, with their guests using the bathing facilities of the Hotel-Baths Ixtapan de la Sal, but almost never the rustic baths at El Bañito and Santa Catarina, which served the poorer, local population. Another entrepreneur, Carlos Rodríguez, arrived in 1933, buying land around the town and investing 100,000 pesos to build the Hotel Casa Blanca.[17] By 1940 there was another hotel, as well as guesthouses run by American and German immigrants that catered to foreign visitors.

The increasing control over the business of bathing exercised by outsiders was paralleled by a shift in the legal status of Ixtapan's springwaters, from municipal property to federal property. Mexico's 1926 irrigation law declared much of the country's waters to be the property of the nation, and initiated a process of centralization by which local actors were required to either apply for confirmation of existing uses of waters or ask the government for a new concession of those now-national waters. Just as it had in Topo Chico, Nuevo León (chapter 7), this process led to confusion and conflict in Ixtapan de la Sal. Ixtapan's hot springwaters were declared national property by President Pascual Ortiz Rubio in 1932 with the assertion that they flowed into the Río Balsas, but neither the municipal government nor Reynoso himself seemed to have been aware of that change, or the requirement that any existing uses of nationalized waters be registered with the secretary of agriculture within the five-year period that followed.[18] Because of the failure to register these existing uses within five years, the municipality of Ixtapan de la Sal lost rights to the springs, and the contract with Reynoso for those rights was thus null and void.[19]

José Vergara was angry. As municipal president, in 1930 he had signed the Municipality of Ixtapan de la Sal to a 25-year contract with José Reynoso, in which the businessman was given almost total control over the waters of the town, in exchange for fifteen percent of the profits generated by those waters. Now only a year later he was sitting in the office of a notary public in Tenancingo, lodging a claim to the bathhouse and the land where the San Gaspar springs bubbled up. The land and springs were "property and possession" of the people of Ixtapan "since forever," testified residents Juan Hernandez and Onofre Morales. Vergara felt cheated by Reynoso, who had convinced him through "tricks and false promises" that the business would provide a healthy income to the municipality. He could see the money flowing into Ixtapan de la Sal, with outsiders buying land and building houses and hotels, but only a pittance for the townspeople. It was time to fight.

Source: AHA, AN, Caja 2058, Exp. 31075 (1939); Escrituras de Propiedad (1933); AHA, AS, Caja 2058, Exp. 31075, pp. 151–55, Vergara to Olivier Ortiz.

The problematic legal status of Ixtapan's hot springs soon became apparent in the context of a struggle between national businessmen and local townspeople to control the benefits of bathing tourism in Ixtapan. In 1933, a year after José Vergara signed the contract with Reynoso, the municipal government registered a claim to the land and buildings of the Hotel-Baths Ixtapan de la Sal with a notary public in Tenancinco. When the five-year window for claiming existing water uses with the federal government closed in 1937, the municipal government asked that the secretary of agriculture recognize its right to waters that "have been used since before anyone can remember as public baths."[20] In 1938 the municipal government filed concession requests for all the springs in Ixtapan, pointing out to the secretary of agriculture and development (Secretaría de Agricultura y Fomento, or SAF) that since 1932 they were national waters and that neither Reynoso, the municipality, nor anyone else had ever filed a request for their concession. The municipality also tried its luck with other branches of government, filing a claim with the Department of Indigenous Affairs, and even writing to the president.[21] The secretary of agriculture and development, charged with managing national waters through its General Directorate of Waters (Dirección General de Aguas), ruled in 1939 that both Reynoso and the municipal government had failed to formalize their claim to the national springwaters of Ixtapan, and recommended that the towns-people form a cooperative to ask for the concession to supply a modern bathhouse.

The presidency of Lázaro Cárdenas (1934–40) was a period of social reform, marked by the nationalization of the oil industry and large landholdings, the strengthening of labor unions, and the creation of collective farms. The SAF, which controlled the concession of waters, was especially supportive of peasant economic initiatives during this period, and so José Vergara and the townspeople of Ixtapan stepped up the pressure on Reynoso, hiring anarchist lawyer and revolutionary precursor Enrique Flores Magón to represent them.[22] They formed a cooperative, and filed a new request for water to supply a community bathhouse, and with the profits, to improve the road to Ixtapan, clean up the town for tourism, and build a school, library, and hospital. At the same time, the municipal government withdrew its request for the waters, in the knowledge that Mexican law at the time privileged cooperatives over others requesting to use natural resources. In a strange irony, the municipal government found itself arguing that it had no right to the waters and that the federal government did (thus invalidating the contract), and Reynoso ended up arguing that the water belonged to the municipality (and therefore his contract was valid). As documentation of the multiple legal issues piled up on the desks of the somewhat bewildered and hesitant SAF officials, the townspeople grew increasingly hostile to Reynoso and his Hotel-Baths.

Faced with the knowledge that his contract with the municipal government of Ixtapan de la Sal was invalid, that the springwaters were national property, that the town was determined to recover them, and that the federal government was generally supportive of such actions, Reynoso looked for a way out. He found it in 1940, through an offer by the company Campos Mexicanos de Turismo (CMT) to transfer title of the Hotel-Balneario Ixtapan de la Sal in exchange for 70,000 pesos of stock in the company.[23] What made the CMT's offer attractive was that its boss was Pascual Ortiz Rubio, ex-president of the Republic (1930–32) and highly connected leader of the growing tourism industry. Ortiz Rubio had clear channels of influence with the federal government, all the way up to the president of Mexico, and on multiple occasions between 1941 and 1945 wrote directly to President Ávila Camacho, or Minister of Agriculture Marte R. Gómez, to advance the interests of the CMT. He was also familiar with Ixtapan de la Sal, for it was he who, as president of the Republic, authorized the nationalization of the town's springwaters in 1932.[24]

Ortiz Rubio created CMT in 1935, pooling money from a group of investors to buy and operate the "Hotel Valles," in Ciudad Valles, San Luis Potosí, a habitual stop for American tourists heading south on the Pan-American Highway.[25] His development vision included hotels for international tourists and the infrastructure to serve them. CMT's project to develop Ixtapan centered on enlarging the existing hotel from 34 to 100 rooms, expanding the bathhouse on the San Gaspar springs, bringing more potable water to the hotel, and convincing the federal government to pay his company to build a paved highway to connect Ixtapan with the urban centers of Toluca, Mexico City, and Cuernavaca and the tourists sites of Metepec and the Grutas of Cacahuamilpa.[26] CMT bought the Hotel-Baths Ixtapan de la Sal in 1940, and Ortiz Rubio soon presented Ávila Camacho with his proposal to rebuild it as a more modern, international hotel that would look like a "European spa." A chemical analysis of the waters of San Gaspar springs showed them to be rich in sulfates, potassium and sodium chlorides, and useful to treat rheumatism, gout, insomnia, and maladies of the nervous system and skin.[27] Echoing the nineteenth-century discourse of the water cure, Ortiz Rubio promised "a sanatorium built with all the required and most up-to-date scientific knowledge."[28] But the focus was on tourism. It was to be, he confided to the president, part of a major development that included the sale of properties for country houses, a highway to Mexico City and Cuernavaca, and a redesigned town layout.

Along with convincing the federal government, Ortiz Rubio and the CMT also had to negotiate with the people of Ixtapan de la Sal, who declared their opposition to the 1930 contract with Reynoso (now with CMT) even before the ink had dried. By 1940 the town organized a cooperative and formulated its own development plan, and Ortiz Rubio, sensing the seriousness of the community's resistance, incorporated some of the elements of this counter plan into a compromise offer. In exchange for disbanding the cooperative, the CMT offered to build a school in the town, and to give a free parcel of land to the municipality for enlarging the bathhouse at San Gaspar. Flores Magón rejected the offer on behalf of the cooperative.

In mid-June 1941, after years of uncertainty, the federal government ruled on the situation. The lawyers of the SAF determined that the Reynoso-municipality contract was valid, but that neither party acted to confirm rights to the water after their nationalization in 1932, and thus neither had rights. Following from this, the municipality had no legal basis by which to pass its rights to the cooperative. The rights to the waters were still the federal government's to award, and President Ávila Camacho ruled, in a Presidential Accord, that the state of affairs would continue, with the CMT enjoying the use of the San Gaspar springs, but under the assumption that it would make a major investment to build a bathhouse and hotel "with all the modern conditions and comforts."[29] Ávila Camacho reminded Ortiz Rubio of the importance of the "hydrotherapeutic center of Ixtapan de la Sal" for the national strategy of tourist development, and told him that if the CMT could not make such an investment, the water rights would be assigned to someone who could "offer the hope of prosperity both for the locality and for the country's tourist industry."[30] Sensing the legitimacy and power of the position the community had established over the previous years, the president also moved to protect existing free access by locals and "the poor and needy" to the other hot springs in Ixtapan—El Bañito, Santa Catarina, and Laguna Verde.

The Presidential Accord spurred both the CMT and the townspeople to present development plans for Ixtapan that heeded its goals. The CMT, emboldened by the accord, filed its plan along with requests for water from San Gaspar, El Salitre,

El Ojito, and the freshwater spring.[31] The cooperative promised much the same: to build the highway, expand the bathhouse, beautify the town for tourists, and build "campgrounds for tourists . . . chalets or bungalows . . . and fields for tennis, basketball, golf, etc." They would publicize the town internationally in Spanish and English, stressing the "curative properties of the waters," and proceeds from tourism would build schools, a theater, a hospital, a library, and other community services. The government reiterated that the waters would be given in concession to the agent who could raise the capital for investment in a first-rate tourist hotel and bathhouse.[32] Flores Magón replied that it was unconstitutional to make the concession of the mineral springwaters to the cooperative conditional on its economic capacity to develop them, and maintained that it had legal precedence for using the waters for development.[33] The state, he argued, had placed its desire for tourist development above the law.

The most important requirement of the Presidential Accord was that whoever controlled the springs would invest large amounts of capital in developing them. The CMT offered ledger sheets and corporate reports as evidence that it could deploy the necessary funds, but the townspeople could only promise the secretary of agriculture that once the water concession was awarded, the cooperative would sell 60,000 pesos of bonds to build the bathhouse. The hotel, they maintained, was not necessary because there were already hotels in Ixtapan. This argument, and the claim that the town could raise the capital, was met with skepticism, and the government rejected the townspeople's petition on the basis that "it did not offer sufficient hope of viability."[34]

Struggling to raise money, the townspeople turned to the next biggest businessman in Ixtapan after the CMT, Carlos Rodríguez, owner of the Hotel Casa Blanca. In 1942 Rodríguez submitted a request for a concession of waters of the Santa Catarina spring, along with a plan to build two swimming pools, a free one that satisfied the Presidential Accord's requirement for access by townspeople and the needy, and another for tourists whom he would charge entry.[35] He presented himself as a man of the people who brought prosperity to the town through his hotel, and offered to invest 500,000 additional pesos in rebuilding that hotel and the new bathhouses at the Santa Catarina spring.[36] The municipality, seeing its chances to control the springs growing slimmer, declared its support for Rodríguez's plan. On the other hand, CMT director Pascual Ortiz Rubio wrote a letter to President Ávila Camacho vehemently protesting the plan for contravening the Presidential Accord, encroaching on the CMT rights established by the 1930 Reynoso-Municipio contract, and promoting unhygienic bathing practices.[37] Ortiz Rubio hoped to persuade the president to order Secretary of Agriculture Marte R. Gómez, an old *agrarista* sympathetic to peasant causes, to reject Rodríguez's development plan and his request for the springwaters, but when this pressure went unheeded, the CMT filed an appeal (*amparo*) in court.[38]

Carlos Rodríguez shifted uneasily in his hard wooden chair in the office of the Directorate of Waters, in the Tacubaya neighborhood on the western outskirts of Mexico City. Rodríguez brought his lawyer to the meeting with Olivier Ortiz, manager of the Campos Mexicanos de Turismo, and officials of the Secretaría de Agricultura y Fomento, but it did not make him feel much better about the situation. He was up against Pascual Ortiz Rubio, former president of Mexico, and he felt outgunned in this struggle over the springwaters of Ixtapan. The "junta de avinencia" that he had come to participate in was supposed to reach a compromise settlement, and Rodríguez hoped it would, to avoid further legal costs and delays in the construction of his planned bathhouse in Ixtapan de la Sal. He had followed the letter of the law, including that of the Presidential Accord, but the Campos Mexicanos de Turismo fiercely opposed any competition for the business of bathing in Ixtapan, and word had it that the Ministry of Agriculture only entertained his development plan and request for waters because the secretary himself, Marte R. Gómez, refused to bend under the pressure applied by President Ávila Camacho in support of Ortiz Rubio. As the CMT lawyer spoke, it became increasingly clear that there was no chance they would move forward with his plans. The CMT had already agreed with the government of Mexico State on a plan to develop Ixtapan, and Rodríguez was not part of it. Worse still, Ortiz accused him of trying to dispossess the townspeople of the Santa Catarina spring; and he knew that the CMT employees were going around town undermining his efforts by telling people that they were "selling their birthright for a plate of lentils." Well, they will likely win in the end, he thought, but he was not going to give in at this meeting.

Source: AHA, AN, Caja 1206, Exp. 16354, pp. 178–95, Transcript of the Junta de Avinencia (October 27, 1943).

At the beginning of 1944, the legal status and control over the hot springs of Ixtapan de la Sal was still not defined. Three years earlier President Ávila Camacho had validated the status quo possession and use of the San Gaspar spring by the CMT, but the SAF had never awarded a definite concession of those waters to anyone. In the midst of this uncertainty, the municipal government denounced to Ávila Camacho and Marte R. Gómez the water-grab attempted by CMT and, in an about-face, rejected Rodríguez's request for the Santa Catarina spring.[39] When the SAF decided to award Rodríguez the concession to the Santa Catarina spring, they made it contingent on the approval of the municipal government, which was not forthcoming. Instead, the municipal government returned to the SAF with its own plan to build a pool at the Santa Catarina spring.[40] In an effort to buy the compliance of the municipal government, the CMT brought its own plan for a municipal bathhouse at the Santa Catarina spring, and offered the land, the engineers, and 500 pesos toward its construction. Much like the Presidential Accord reached in 1941, the SAF ruling in 1944 left the situation in Ixtapan unresolved.[41]

A solution, however, was already being negotiated between the municipal government and Arturo San Román, a businessman with deep roots in Ixtapan. San Román enjoyed a certain measure of local trust because his father, Atilano San Román, had arrived in the town in 1890. More important, however, is that he was also able to muster the capital required to turn Ixtapan into a tourist destination, and his connections to state and federal governments helped him push through decisions.[42] San Román's vision for Ixtapan went far beyond rebuilding the Hotel-Balneario, and included the creation of a country club–style development outside of town known as Nueva Ixtapan, with homes built around a new reservoir. To support this project, he planned to ask for concessions not only of the hot springs, but of the freshwater from the canal that supplied the town and its agriculture.[43]

San Román was able to work with the CMT and municipal, state, and federal governments to line up all the pieces of his development plan in the space of a few years. In August of 1944 Ortiz Rubio wrote his friend President Ávila Camacho to ask him to modify his Presidential Accord so that the CMT and San Román could share the project to build a first-rate hotel and *balneario*, with CMT rebuilding the hotel and San Román investing in the bathhouse and processing the water concession.[44] In September San Román asked the municipality to sell him lands on which the San Gaspar Balneario was located, and in October the CMT sold San Román its water rights to the San Gaspar, El Ojito, and El Salitre springs. Still, the municipal government and San Román remained at odds over a number of issues, so on December 21, 1944, they presented their arguments for arbitration by the governor of the state of Mexico, Isidro Fabela.

Fabela's ruling was accepted by all parties. They agreed to revise the payment for the land and waters of San Gaspar that had been stipulated in the 1930 pact between Reynoso and the municipal government. San Román was required to pay for and build two *balnearios*, a low-cost one for locals at the Santa Catarina spring, and a free one at the El Bañito spring for the needy. Income from the Santa Catarina baths would be administered by an independent board, and used to maintain both baths and make other material improvements in the town. The municipality would, furthermore, provide land to San Roman outside of town for a dam to store water from the freshwater canal to be used by the town, and cede to him the property rights of the parcels of land upon which the San Gaspar springs and its hotel and *balneario* were located. The transactions compelled by the state's arbitration established the definitive rights to land and water that paved the way for capital investment in Ixtapan de la Sal.

Springwaters were a crucial ingredient of the development project in Ixtapan orchestrated by San Román, but so too was freshwater. The nineteenth-century canal that brought water to the fields and homes of Ixtapan was insufficient to supply the growth of San Román's New Ixtapan complex, so he worked with

FIGURE 20. View of the Nueva Ixtapan, S.A. housing development, c. 1955. With permission of the Archivo Histórico del Agua, Mexico City, Mexico. AHA, AN, Caja 1204, Exp. 16337.

state and federal governments to improve the infrastructure and increase supply. As early as April of 1944—long before the arbitration ruling was delivered by Governor Fabela—San Román presented the SAF with plans for the construction of hydraulic infrastructure that would convey and store water from the Barranca del Calderón, and convinced Augusto Hinojosa, Mexico State's senator to the National Congress, to order a study of the freshwater available in that barranca and the eighty-kilometer canal.[45] As soon as the arbitration decision was delivered by Fabela, the SAF sent an engineer to identify water that could be brought to Ixtapan, and once it was determined that there was an available amount of 238 liters per second, San Román filed a request for that amount. He also asked to make temporary use of 1,000 liters per second of the flow of the Barranca to drive electricity-generating equipment.[46] The rest of the towns along the canal were told that gates would be installed to limit their flows of water, and in 1946 San Román received a fifty-year concession from the SAF to provide potable water to his new tourist development.[47]

The arbitration decision resolved the long struggle between the municipal government and outside developers over the San Gaspar springs, and satisfied the terms of Ávila Camacho's Presidential Accord by channeling capital into the creation of a center of international tourism in Ixtapan de la Sal. The fate of Ixtapan's other springs remained unclear, however. San Román offered to build two bathhouses on the El Bañito and Santa Catarina springs, but Carlos Rodríguez already held the conditional concession of these waters from the SAF. Once San Román

FIGURE 21. Municipal bathhouse, Ixtapan de la Sal, c. 1965. Courtesy of Luis René Arizmendi, Ixtapan de la Sal, Mexico.

took charge of the development of Ixtapan, the SAF attempted to take that concession away, but Rodríguez lodged a legal appeal and Mexico's Supreme Court ruled that the concession was valid even without the permission of the municipal government.[48]

"In reality, the town itself has denied, with weapons drawn, the efforts by Mr. Rodríguez to take possession of the spring. The feelings about this issue are so extreme that I feel Mr. Rodríguez's life would be in danger if he continues any further. . . . The town is not willing, for any reason, to sell or give land to Mr. Rodríguez, and the result is that Mr. Rodríguez has no property on which to build his planned bathhouse."

—Victor Peredo, federal government official, 1953

Source: AHA, AN, Caja 1206, Exp. 16354 (December 4, 1953), Memorandum, Victor Peredo.

Years of conflict ensued, as Rodríguez attempted to move forward with his plan to build the bathhouses against the sustained opposition of the townspeople and the municipal government.[49] He excavated the pools in late 1946, but in March of 1947 construction was interrupted by "various townspeople," and his workers refused to resume their labors. Rodríguez accused José Vergara of scheming with "others from the town interested in obtaining the concession" and petitioned the municipal, state, and federal governments for protection. Witnesses testified that in February of 1950 a group including the municipal president and some police officers stopped the work "at gunpoint."

Rodríguez received no response to these requests for protection, and the works remained abandoned until 1951, when the SAF finally revoked his water concession and building permit.[50] The conflicts between Rodríguez, San Roman, and the townspeople continued to simmer, and 1954 the municipal government passed a resolution to prohibit access by Rodríguez and his workers to the municipal lands on which the springs and bathhouse project were located.[51] Rodríguez finally gave up, and in the late 1950s the municipality and townspeople secured the concession of the El Bañito and Santa Catarina springs and built the municipal bathhouse foreseen in the arbitration decision of 1945.

CONCLUSIONS

The deal reached between San Román and the municipal government of Ixtapan de la Sal enabled politicians and businessmen to develop the town as a tourist destination, and secured some benefits from that development for the townspeople by providing them with a share of the business of bathing through the construction of a municipal bathhouse. In this way, the long struggle of the townspeople to maintain control or access over their local mineral springs was successful. Nevertheless, the townspeople participated in the economic development of Ixtapan in a clearly subaltern position. Once the land and water rights were established, San Román was able to organize, with the support of his allies in the state and federal governments, an elite tourist development scheme that utilized the waters and labor of locals to generate profits for him and his family.[52] When the canal was finally opened in 1960, all the towns in the region protested the loss of water to Nueva Ixtapan and asked for their previous concessions to be restored.[53] By this time, however, the spa tourism model promoted by the postrevolutionary government since the 1920s was a reality, and tourism in Ixtapan sucked the region's land, labor, and waters into its orbit.

The strategy of placing mineral springs tourism at the heart of economic development has been successful across Mexico, and owes its success to the persistence of strong cultural values for virtuous waters. In the town of Oaxtepec, Morelos, for example, an enormous vacation center and water park was built in the 1960s by the federal government's Mexican Institute of Social Security (IMSS)—the Centro

Vacacional de Oaxtepec. Oaxtepec's hot springs are only one of many hydraulic features of the Centro Vacacional, which sports an Olympic-size swimming pool, a diving pool with platforms up to ten meters high, various wading pools, sports fields, and numerous restaurants, bars, hotels, and cabins. In 2017 the range of water-based activities expanded even further, as the Six Flags Company opened its "Hurricane Harbor" theme park on this public property. Nearby, in the city of Cuautla, Morelos, the hot springs of "Agua Hedionda" have been developed into a vacation destination, also with a wide range of activities. The state of Hidalgo is famous for its hot springs bathing complexes. And so on, throughout Mexico.

During the twentieth century, hot springs throughout the country were converted into *balnearios* run by *ejidos*, town governments, the federal government, and private businesses. The National Water Archives in Mexico (Archivo del Agua) holds hundreds of pages of documents concerning dozens of hot spring during this period, when plans were drawn up to turn them into tourist centers, or perhaps to develop bottling plants using their waters. In each setting a different bargain was struck between local inhabitants and outside businessmen eager to cash in on these virtuous waters. It certainly matters who ends up controlling these waters and the businesses that make use of them, as the sheaves of archival documents chronicling struggles over mineral springs can attest. But regardless of the particular outcome of primitive accumulation, bottling, and spa tourism, the ongoing popularity of these heterogeneous waters is evidence of the deep attraction they continue to exercise. It would be impossible to understand the political ecology of mineral springs without considering the virtues of these waters.

9

Virtuous Waters in the Twenty-First
Century

Every weekend people from Toluca, Mexico City, and other nearby areas pour into
Ixtapan de la Sal to take the waters. Those who can afford it go for a few days of
relaxation around a pool at a fancy hotel or private home. Many families will visit
the Ixtapan Aquatic Park, a somewhat-pricey waterpark with slides, rides, and restaurants, but it is mostly the old folks who head for a soak at the park's hot springs
bathhouse. The older people remember when the bathhouse was connected to the
hotel, before the San Román family divided their property and the aquatic park
was built to appeal to younger generations of customers seeking more exciting
interactions with Ixtapan's waters. The aquatic park also has a luxurious spa with
thermal waters to serve upscale guests. Visitors to the town with more modest
resources perhaps choose a smaller hotel with a pool, and those with less will stay
at a budget hotel or pension in town and walk to the municipal bathhouse, which
is about a third of the price of the waterpark, but still offers various pools, a water-
slide, and a restaurant.

Many of those at the municipal bathhouse come to Ixtapan to soak in the therapeutic hot springwaters. "The waters are salty," a middle-aged man told me as we
lounged in one of the two small hot pools at the municipal bathhouse, "not like the
sulfurous waters of Hidalgo. These ones are good for reducing blood pressure." The
small hot pools were considered therapeutic by most everyone sitting in them, and
the cooler large pool at the center of the complex was more for fun, as evidenced
by the kids, babies, and parents splashing around. An older woman said that the
salty waters of the hot pool were good for her rheumatism, and she moved to let
the water flow over her as it emerged from an opening in the floor at the center of
the pool. I hauled myself from the hot pool to the murky, cool waters of the nearby

FIGURE 22. A weekend afternoon in the Municipal Bathhouse, Ixtapan de la Sal, 2009. Photograph by author.

mud pool, where a few men and women sat plastered with the dark brown mud of the hot springs. "It removes toxins," an overweight young man told me as we sat with the mud drying on our faces, looking a little like aquatic raccoons. A pair of well-dressed young women had set up a table nearby, looking out of place in their high heels. They were selling some sort of dietary supplement concocted from *nopal* cactus, a traditional food in highland Mexico reputed to alleviate diabetes.

I followed a busload of elderly folks as they filed into the bathhouse and entered a separate hydrotherapy room, guided by a younger woman in white pants and jacket who looked and acted like a nurse, giving instructions about how best to take advantage of the healing powers of the Jacuzzi jets and waters. The elderly patients were taking a day trip to treat assorted complaints: mostly high blood pressure, aching joints, and obesity. I soon found out that access to this room required an additional fee, and it was clearly set apart in its ambience from the ludic space of the bathhouse's main pools. The hydrotherapy techniques were prescribed and required more control. Like the hot spring pools it was a healing space, but the bathing practices were regimented and it had a more clinical feel.

Heterogeneous waters continue to attract people to their virtues, but the knowledge and use of these waters is not as common, perhaps, as it once was. Ixtapan

FIGURE 23. Hydrotherapy room, Municipal Bathhouse, Ixtapan de la Sal, 2009. Photograph by author.

de la Sal might be the best-known spa town in Mexico. It is close to Mexico City and continues to thrive as a hub of tourism, mostly national. Ixtapan's waters still attract many visitors, but the great majority of today's visitors to the town will not bathe in the mineral waters that made it famous. The number of hot springs has not increased, while the numbers of hotels, pools, spas, and visitors have sky-rocketed over the last century to accommodate growing numbers of tourists. The upscale guests at the Hotel Ixtapan, for example, might choose an aromatherapy massage at the new "Holistic Spa," or maybe take a yoga class or sunbathe by the pool. Those still interested in soaking in the mineral waters will have to leave that hotel through a gate to get to the old bathhouse, now on the property of the neighboring Ixtapan Aquatic Park, or they may use the individualized bathing rooms of the marble-and-onyx-covered Ixtapan Spa, built in the 1960s. The old bathhouse is something of a relic, "rubble" of a bygone moment.[1] The Ixtapan Spa from the 1960s is dated in its appearance, and its mineral baths do not, the hotel recognizes, reflect "new health and fitness trends" such as those offered at the "Holistic Spa".[2] Some other boutique hotels in Ixtapan have built *temazcal* steam-baths for guests seeking the curative and spiritual dimensions of pre-Hispanic and indigenous bathing practices. Ixtapan's mineral springs still bubble from the ground, but most people choose a bathing experience in which mineral waters do not play a central role.

Visitors to other historically famous mineral springs in Mexico encounter a similar situation. Little remains of the bathhouse row in Aguascalientes, and in Tehuacán the efficacious, virtuous waters that drew people to the town for centuries are rather hard to find these days. The source of the Peñafiel mineral waters can be seen but not touched or even photographed, deep in an underground cavern below the bottling plant owned now by the Dr. Pepper–Snapple group. The El Riego and Garci-Crespo hotels are long gone, and the hotels that serve visitors do not offer mineral waters for bathing; they might offer other present-day "spa" services such as massages, face masks, or chocolate "therapy."[3] The mineral waters of Tehuacán are now only accessible for bathing at the San Lorenzo spring, which supplies a shady, popular waterpark two kilometers from the town center. At San Lorenzo, the communitarian form of *ejido* property has conserved the springs as a common-pool resource and has served as a barrier to primitive accumulation. Topo Chico, on the other hand, is now an industrial zone of the city of Monterrey, and the only hint that hot springs once sprang forth there is a private museum dedicated to the history of the Topo Chico mineral water company. There are plenty of spas advertised in Monterrey—"nail" spas, "day" spas, "medical" spas—but none of these involve bathing, drinking, or otherwise interacting with mineral waters.

The mineral spring in the "El Pocito" church in La Villa de Guadalupe can be smelled and not seen: its sulfurous scent wafts up from a well that remains grated to prevent anyone from interacting with its waters. "El Pocito" is now known for its architecture rather than for the fact that it enshrines a centuries-old engagement with the heterogeneous waters of the Valley of Mexico. The nineteenth-century bathhouse founded by Liceaga in front of the train station in La Villa de Guadalupe is long vanished, and the springwaters that served as a pilgrimage destination for centuries are only symbolically present in the elaborate fountains, pools, and waterfalls that have been built into the side of the Tepeyac hill that rises above the Basilica of Guadalupe and its uber-iconic depiction of the Virgen de Guadalupe. These waters are off-limits to visitors, who will also find no holy water offered at the entrance of the various churches at the site. But traditions are slow to fade, and every day dozens of young parents bring their babies to be baptized in the "Bautisterio" (baptismal chapel), a large building next to the basilica dedicated solely to that important watery sacrament.

In Peñón de los Baños, the thermal mineral waters have simply been neglected as capital chases better chances of accumulation. But because they are not used for bottling or spa tourism, the therapeutic uses of the waters stand out. The bathhouse at Peñón de los Baños is today located on the dark, decrepit, and austere bottom floor of a brick and cement building in a drab working-class neighborhood in Mexico City. The modest business serves folks who seek a medical treatment; there is no luxury, no leisure. Instead there are private rooms with *placeres* and hot

FIGURE 24. El Pocito, Villa de Guadalupe, 2017. Photograph by author.

water drawn from a well—the springs, exhausted by groundwater pumping, no longer bubble to the surface (see Figure 1). Everyone who goes there is convinced of the medicinal efficacy of bathing in the waters, but few others even know about them, least of all doctors. Like the tourism industry, the medical profession no longer depends on mineral waters such as those of Peñón or Ixtapan; it seems that most doctors share the view of water as homogeneous and inert, good for washing.

Despite the general decline of mineral spring bathing during the twentieth century, heterogeneous waters still exist, and people are still drawn to them for cultural reasons that have accumulated over centuries. Drinking is the most common interaction with mineral waters today, and it is clearly enjoying an immense upwelling of popularity. In a few cases—Tehuacán, Peñafiel, and Topo Chico—the mineral springs have been captured for bottling, and these waters are indeed well known. Like all bottled waters today, mineral waters appeal to consumers because of their specificity—the fact that they are not tap water. Bottled waters are felt to be purer and safer than tap water, and those that claim an attractive origin—"bottled at the source"—are especially desired. The "Tehuacán" brand, for example, says that its water originates at the top of the Pico de Orizaba, a nearby volcano.[4] Mineral waters such as Topo Chico, Peñafiel, and Tehuacán carry these connotations of purity derived from geographical specificity, but their perceived virtues go beyond purity to include the efficacy of their mineral contents. Mineral water is seen to help digestion, and the chemical contents of the waters are sometimes displayed on the label as if they were medicine. Long-standing ideas about the influence of the environment on human health animate the ingestion of bottled mineral waters, and mineral waters remain virtuous and efficacious in the minds of many consumers, in Mexico and around the world.

Bathing is now a less-common form of interaction with mineral waters, but many hot springs remain popular as mass tourist destinations, and long-standing concepts of the healthfulness of immersion in these waters are still held by those who visit them. I spent many afternoons, for example, in the hot spring pools of the Balneario Municipal in Ixtapan de la Sal, chatting with the visitors. Sitting together in the hot pools, we would discuss the virtues of the waters, and how they acted upon the body to ease problems of gout, arthritis, high blood pressure, circulation, etc. The salt in the water was good for allergies; the dissolved carbonate gas soothed the nerves. The mud we smeared on faces, arms, and bodies kept our skin smooth and youthful. Scientific proof of the efficacy of the waters was in plain sight: hung on the walls were analyses of their chemical composition. Also common, however, were stories of miraculous cures. One woman told of a deep scar that was made invisible by the waters; another witnessed a man enter in a wheelchair and leave walking. Bathing in these waters was a deeply social activity, sitting elbow to elbow in the water and talking about how it soothed our maladies.

What can we learn from these waters and cultures? What is their place in the twenty-first century? I have approached this complex multifaceted history from an historical, anthropological perspective that views human relations with the environment in a holistic way, informed by political ecology and its emphasis on social organization, conflict, and materiality. The history of waters and cultures in Mexico shows complex dialectics: heterogeneity and homogeneity; curing and contagion; social and individual; and a subject/object dynamic that posits water as both an inert substance to be controlled by humans, and as an efficacious, agential substance that works changes on humans. One of the trends in this history is toward the expansion of infrastructure and an associated individualization of our contact with water. Bathing for cleanliness has moved from public spaces to private ones. Shared bathhouses have given way to private bathrooms in people's dwellings. But this is not a unilineal process of water displacing waters; homogeneity has not abolished heterogeneity. Rather, the effort to create a sanitized, uniform, public water during the nineteenth and twentieth centuries actually strengthened the identities and values of specific waters, and we see that social bathing for fun, fitness, and therapy continued to flourish alongside the shower. Sit in any hot spring in Mexico, and this is evident.

Waters define social groups and their boundaries. Despite the individualization of bathing, our engagements with water are still organized around group identities. Soaking in a small pool of water with dozens of perfect strangers is something that many people are uncomfortable with, because intimacy probes the borders of race, class, ethnicity, sex, and gender that our society is built upon. Wealthy clients of Mexico City's bathhouses in the 1790s bathed in private *placeres,* away from the *temazcales* and shared waters of the *plebe.* A century later the rich bathed in the privacy of their own homes in order to escape the massification of bathing. Social and sexual intimacy in the *temazcales* of New Spain was enough to engender a full-scale colonial clampdown; sex in the baths during the Porfiriato was greeted with a similar reaction. Boundary work is also busy at the borders of race and class. In Ixtapan de la Sal, the municipal bathhouse was built to cater to locals and people of more humble economic means; the hot springs at the Hotel Ixtapan served the wealthy. Today, the town's hot springs serve a wide cross-section of society, but those with a lot of money retire to more exclusive spas that have no mineral waters at all.

Along with the social distinctions that organize social practices of bathing, an important dynamic in the history of heterogeneous waters is the ebb and flow of interest by capital in them. The bathhouse at Peñón was renovated and served wealthy clientele in the late eighteenth century, and then again in the late nineteenth century. All along, however, humble Mexicans made use of the waters, pushing back against businessmen and governments in order to retain access to the common resource. Hydraulic opulence in the late-nineteenth century fomented the business of bathing and made these waters accessible to many more Mexicans. In Spain and other areas of Europe, elite interest in mineral waters declined during the mid-twentieth century, when public health systems enabled wide access by the middle class and even poor. With the decline of the welfare state, elite spas are once again on the rise, as these commons are privatized and investors eye the business of bathing as an attractive way to tap the wealth of the rich. Throughout Mexico, however, access to mineral springs remains fairly open, conserved that way in many cases by municipal and *ejidal* ownership and administration. It remains to be seen if another wave of primitive accumulation will turn the popular spas of Mexico into luxury resorts, but the strength and persistence of Mexican water cultures and their traditions of access make such an outcome seem unlikely.

Cultures of waters deserve our attention. Most of what we know about our relation to water has to do with the macroenvironmental aspects of irrigated agriculture and urban water systems. But waters move through our bodies and intimate social spaces as well, and scholars have largely ignored these topics. As the world confronts a generalized water crisis, intimate engagements with waters will surely play a role in reconstructing a more sustainable relationship with the liquid. Already the strategy of increasing supply through the construction of ever-expanding storage and conveyance infrastructures has been rejected in favor of demand management, which pivots around reducing domestic water use in urban contexts. More efficient bathing practices are in order; shorter, less frequent, perhaps shared engagements with water. Ancestral water cultures present a range of affective values that may point us toward less destructive uses. We fill our bathroom drains and public water infrastructures with soap and shampoo, but we might not be comfortable about dumping these substances in a pond that we drink from and swim in. A return to feeling and valuing the specificity of heterogeneous waters and promoting of inter-actions with them could limit the destructive aspects of the modernist illusion of unlimited, homogeneous water disconnected from our bodies and lives.

Virtuous waters, past and present, convey elements of more sustainable rela-tionships with the environment and with each other. Unfortunately, building a sustainable relationship to water is not simply a question of replacing one cultural element with another. Heterogeneity, for example, contributes content to our dys-functional and maladaptive relationship with water, as evidenced by the recent boom of bottled waters. These waters are sold as unique and geographically spe-cific: sometimes they are, often they are not. Regardless, the production of billions of plastic water bottles drives up costs for the liquid, pollutes our rivers and oceans, and generates fabulous profits for the few while defunding the public water sys-tems that do the same job much more efficiently.[5] Heterogeneity lies at the heart of the commodity form, and this complicity promises to cause mischief as we chart a course toward more local, more intimate engagements with the hydrosphere.

Renovating our relationship with our waters requires a focus on, well, relations: with water but also with each other. The search for profit that lies at the heart of our economic system will relentlessly define the liquid by the part it plays in the accumulation process. It may be a resource that provides rents through irri-gated agriculture, or it can generate profits through bottling and sale as a com-modity. However, the quest for profit propels the business of bathing and bottling unevenly, in cycles and surges, never fully eradicating the relations of community and cooperation that ensure access to springs at places such as Peñón and Ixtapan, or erasing deep-seated ideas about the virtues of the liquid. The control of mineral waters by some people through law, science, technology, and sometimes outright force was and still is resisted with some success by other groups of people. As we struggle toward a more sustainable relationship to waters, we may find that these deep, long-standing relationships that everyday people have with waters and with each other can serve as a fountain of ideas and practices for building more virtu-ous cultures of waters.

NOTES

1 WATERS/CULTURES

1. Jackson 1990.
2. Johanson 1997.
3. Brewer 2004; Girón Irueste 2006.
4. Kauffman 1959.
5. Mora 1992.
6. Harley 1990; Palmer 1990.
7. Archibald 2012; Blásquez and García-Gelabert 1992; Olmos 1992.
8. Corbin 1988; Gandy 2002.
9. Coley 1979, 1982; Hamlin 1990; Porter 1990.
10. Anderson and Tabb 2002; Brockliss 1990; Chambers 2002; Gil de Arriba 2000; Mackaman 1998; Lempa 2002; Mansen 1998.
11. Brockliss 1990; Cayleff 1987; Dávalos 1997; Goubert 1989; Harley 1990; Melosi 1999; Rodríguez Sánchez 2006; Rodríguez 2000.
12. Bacon 1997; Jarrasse 2002; Walton 2012.
13. Jennings 2006; Lopes Brenner 2005; Quintela 2004.
14. Corak and Ateljevic 2007.
15. Goubert 1989; Vigarello 1988.
16. Corbin 1994.
17. Wiltse 2007.
18. Gandy 2002, 2014; Melosi 1999.
19. Clark 1994.
20. Erfurt-Cooper and Cooper 2009; Cátedra Tomás 2009.
21. Hamlin 1990.
22. Hamlin 2000; Linton 2010.
23. Linton 2010: 81–89.

24. Reisner 1993.

25. Palerm 1973; Steward et al. 1955; Steward 1949; see Walsh 2012 for a discussion.

26. Adams 1966; Geertz 1973; Lansing 1987; Hauser-Schaublin 2003.

27. Aboites 1987; Pisani 2002; Walsh 2008; Ward 2003; Wolfe 2017; Worster 1985.

28. Gelles 1994; Guillet 1992; Hunt 1988; Mabry 1996; Mitchell 1976.

29. Hunt et al. 1976; Lansing 1987; Trawick 2003.

30. Gleick 2010; Kaplan 2008; Wilk 2006.

31. Gleick 2003; Reisner 1993; Zetland 2009.

32. Melville and Whiteford 2002.

33. Walsh 2011.

34. Arrojo 2006; Martínez Gil 1996.

35. Arrojo 2006; Aguilera Klink 1998, 2006.

36. Strang 2004.

37. Wiltse 2007.

38. Jennings 2006; Blakeslee 2010.

39. Kaplan 2008; Wilk 2006; Gleick 2010.

40. Gandy 2104; Linton 2010.

41. Bennett 2010; Latour 2005; Wagner 2013.

42. Wolf 1972; Greenberg and Park 1994; Steward et al. 1955; Steward 1949; Wolf and Palerm 1955; Escobar 1999.

43. Antonio Gramsci's reflections on class and culture shape this view (Gramsci 1971; Crehan 2002, 2016).

2   BATHING AND DOMINATION IN THE EARLY
MODERN ATLANTIC WORLD

1. Birrichaga Gardida 2004.

2. Roberto Melville, personal communication, 2003.

3. García Sánchez 2004; Tortolero 2000.

4. Rojas 1988, 1993.

5. Palerm 1973.

6. Palerm 1973.

7. Pineda Mendoza 2000: 21–55.

8. Birrichaga Gardida 2004: 94–95.

9. Gibson 1964; González Jacome 2007; Palerm 1955; Wolf and Palerm 1955.

10. García Sánchez 2004.

11. Alcina Franch 1994; López Austin 2012.

12. Noriega Hernandez 2004.

13. Schendel 1968.

14. Cervantes Salazar 1985, quoted in Birrichaga Gardida 2004: 294.

15. Schendel 1968: 39–43.

16. Bonfil 1987.

17. Miller 1962.

18. Pliny 1948.

19. Casal García and González Soutelo 2010: 31.

20. Williams 1977.
21. Lara García 1997: 23.
22. Brewer 2004: 185.
23. Girón Irueste 2006.
24. Girón Irueste 2006: 84.
25. Girón Irueste 2006: 82.
26. Jennings 2006: 170–73.
27. Ruíz Somavilla 1992, 1993.
28. Lara García 1997: 22.
29. Ruíz Somavilla 1992: 172.
30. Ruíz Somavilla 1993: 168–71.
31. Anderson 2002: 114.
32. Ruíz Somavilla 1993: 66.
33. Cited in Girón Irueste 2006: 94.
34. Ruíz Somavilla 1993.
35. Ruíz Somavilla 1992.
36. Vigarello 1988.
37. Porter 1990.
38. Ruíz Somavilla 2011: 4.
39. Theimer-Sachse 2000: 205.
40. López Austin 1985: 36.
41. De Landa [1566] 1973: 39.
42. De Landa [1566] 1973: 39.
43. Rodríguez Rivera 1945.
44. Cárdenas [1591] 1945: 71.
45. Avaleyra Arroyo de Anda 2005; Morales Padrón 1949.
46. Vovides, Linares, and Bye 2010.
47. García Samper 1997.
48. Morales Padrón 1949.
49. Hodgson 2004; Aveleyra Arroyo de Anda 2005.
50. Sauza Vega 2008.
51. De Landa [1566] 1973: 55.
52. Henshaw 1910.
53. Cresson 1938; Mason 1935.
54. Romero Contreras 2001: 134.
55. Cresson 1938; Alcina Franch et al. 1980; Houston 1996; Matarredona Desantes 2014; Noriega Hernandez 2004; Ruíz Somavilla 2011; Virkki 1962.
56. Romero Contreras 2001.
57. Alcina Franch et al. 1980.
58. Cited in Romero Contreras 2001: 134.
59. Cited in Romero Contreras 2001: 134.
60. Cited in Romero Contreras 2001: 134.
61. Groark 1997: 16.
62. Silva Prada 2002; Alberro 1992.
63. Rodríguez 2000.

64. Lanning 1983.
65. Gruzinski 1985: 277–78.
66. Silva Prada 2002.
67. Silva Prada 2002: 49.
68. Silva Prada 2002: 12.
69. Candiani 2014.
70. Musset 1992: 14.
71. Birrichaga Gardida 2004.
72. Pineda Mendoza 2000.

3   POLICING WATERS AND BATHS IN EIGHTEENTH-CENTURY MEXICO CITY

1. Foucault 2009.
2. Candiani 2012 revives a notion of class based in modes of production. I prefer a relational idea of class: Gramsci 1971; Thompson 1978.
3. Endfield 2012.
4. Archivo Histórico de la Ciudad de Mexico (AHCM), Ayuntamiento, Aguas, Documentos Diversos para el Arreglo del Ramo, Vol. 29, Exp. 22.
5. AHCM, Ayuntamiento, Aguas, Documentos Diversos para el Arreglo del Ramo, Vol. 29, Exp. 22.
6. AHCM, Ayuntamiento, Aguas, Documentos Diversos para el Arreglo del Ramo, Vol. 29, Exp. 21, "Relación."
7. AHCM, Ayuntamiento, Aguas, Documentos Diversos para el Arreglo del Ramo, Vol. 29, Exp. 22, "Declaración."
8. *Gazeta de México* 49 (September 18, 1795): 409, cited in Martínez Reguera 1892: 287.
9. Rodríguez 2000: 121–40.
10. Rodríguez 2000: 134–35.
11. Rodríguez 2000: 137.
12. Alzate Ramirez [1792] 1894: 29.
13. AHCM, Ayuntamiento, Aguas, Documentos Diversos para el Arreglo del Ramo, Vol. 29, Exp. 22.
14. *Bando de Revillagigedo* 1793. Copy of the broadside consulted in the Archivo Histórico de la Escuela Nacional de Medicina, Mexico City, Mexico. The text was also published in the *Gazeta de México* (August 31, 1793): 428–32.
15. AHCM, Ayuntamiento, Policía, Baños y Lavaderos, Vol. 3621, Exp. 6 (September 17, 1793): Fray Juan Galindo to Policía.
16. AHCM, Ayuntamiento, Policía, Baños y Lavaderos, Vol. 3621, Exp. 12, "Diligencias practicadas"; De Valle-Arizpe 1949: 397–400.
17. AHCM, Ayuntamiento, Policía, Baños y Lavaderos, Vol. 3621, Exp. 1 (October 17, 1778).
18. *Bando de Revillagigedo* 1793.
19. Rodríguez 2000.
20. AHCM, Ayuntamiento, Policía, Baños y Lavaderos, Vol. 3621, Exp. 12, "Diligencias practicadas. . ."
21. Ruíz Somavilla 2011; Aveleyra Arroyo de Anda 2005.

22. AHCM, Ayuntamiento, Policía, Baños y Lavaderos, Vol. 3621, Exp. 3 (June 8, 1792).

23. Johnson and Lipsett-Rivera 1998.

24. Carreón Nieto 1999: 126.

25. AHCM, Ayuntamiento, Policía, Baños y Lavaderos, Vol. 3621, Exp. 23 (1814).

26. *Bando de Revillagigedo* 1793.

27. AHCM, Ayuntamiento, Aguas, Documentos Diversos para el Arreglo del Ramo, Vol. 29, Exp. 22, "Declaración."

28. Gandy 2014; Vigarello 1988.

29. AHCM, Policía, Baños y Lavaderos, Vol. 3621, Exp. 5 (1793), "Causa formada..." In dramatizing this scene I have used a bit of interpretive license.

30. AHCM, Policía, Baños y Lavaderos, Vol. 3621, Exp. 5 (1793), "Causa formada..."

31. AHCM, Policía, Baños y Lavaderos, Vol. 3621, Exp. 12 (November 20, 1750).

32. AHCM, Policía en General, Vol. 3621, Exp. 21 (1809). A bathhouse on the Calle de la Canoa run by Padres del Carmen was found guilty—by failing to provide a toilet—of forcing washerwomen to defecate in its public space, a doubly troubling act involving the aversion to nudity and to excrement.

33. AHCM, Policía en General, Vol. 3621, Exp. 14 (1797).

34. AHCM, Policía en General, Vol. 3621, Exp. 20 (1804).

35. Pulido Esteva 2011.

36. The dictums discussed by Hira de Gortari Rabiela (2002) are: *Ordenanza de la división de la Nobilísima ciudad de México en cuarteles, creación de los alcaldes de ellos y reglas de su gobierno* (1782); *De Enfermedades políticas de la Nueva España que padece la capital de esta Nueva España en casi todos los cuerpos de que se compone y remedios que se le deben aplicar para su curación si se quiere que sea útil al Rey y al público* (1785); and *Discurso sobre la policía de México. Reflexiones y apuntes sobre varios objetos que interesan la salud pública y la policía particular de esta ciudad de México, si se adaptasen las providencias o remedios correspondientes* (1788). The 1782 and 1788 documents were generated by the Oidor Ladrón de Guevara; the 1785 treatise was written by Hipólito Villoroel.

37. Miller 1962.

38. Agostoni 2003: 3–5.

39. Porter 1990; Jennings 2006.

40. Porter 1990; Hamlin 1990.

4   ENLIGHTENMENT SCIENCE OF MINERAL SPRINGS

1. Wagner (2013) reflects the growing importance to the anthropology of water of Bennett (2010), Latour (2005), and other proponents of ontological and new materialist approaches.

2. Solano 1988.

3. Apenes 1944.

4. Ewald 1985.

5. Olivera 1988.

6. Olivera 1988.

7. Olivera 1988: 76.

8. Hamlin 1990; Linton 2010.

9. Aceves Pastrana 1993, 1996.
10. Beaumont 1772: 7.
11. Beaumont 1772: 54.
12. Beaumont 1772: 99.
13. Beaumont 1772: 102.
14. Beaumont 1772: 104.
15. Beaumont 1772: 27.
16. Beaumont 1772: 95.
17. Beaumont 1772: 95.
18. Beaumont 1772: 96.
19. Beaumont 1772: 9.
20. Margadant 1986.
21. Carreón Nieto 1999: 117
22. Carreón Nieto 1999: 117.
23. *Gazeta de México* 3, no. 1 (January 8, 1788).
24. *Gazeta de México,* no. 23 (November 17, 1784): 192.
25. "Valladolid," *Gazeta de México* 4, no. 22: 205–9.
26. "Valladolid," *Gazeta de México* 4, no. 22: 206.
27. "Valladolid," *Gazeta de México* 4, no. 22: 206
28. Aceves Pastrana 1993: 90.
29. Aceves Pastraña 1993: 90.
30. Cervantes de Salazar 1554, cited in Aveleyra Arroyo de Anda 2005: 42.
31. Aveleyra Arroyo de Anda 2005: 43.
32. Aveleyra Arroyo de Anda 2005: 44.
33. Morales Padrón 1949: 703.
34. Aveleyra Arroyo de Anda 2005: 59.
35. Dumont and de Torres 1762: 9.
36. *Gazeta de México* 6, no. 78 (November 19, 1794): 645.
37. Romero-Huesca and Ramírez Bollas 2003; Zedillo 1984.
38. Rodríguez 2000: 172.
39. *Gazeta de México* 6, no. 79 (November 19, 1794): 654.
40. Ocampo 1794.
41. Cited in Aveleyra Arroyo de Anda 2005: 54.
42. Cited in Aveleyra Arroyo de Anda 2005: 54.
43. Aveleyra Arroyo de Anda 2005: 58.
44. Ocampo 1794: 656.
45. AHCM, Ayuntamiento, Policía en General, Vol. 3630, Exp. 218 (April 3, 1827): Jose Maria Manero to Juez Policía.

5   GROUNDWATER AND HYDRAULIC OPULENCE IN THE LATE NINETEENTH CENTURY

1. García Cubas 1904; De Valle-Arizpe 1949.
2. Rivera Cambas 1880–1883, vol. 1: 284–98.
3. Geels 2005.

4. Wiltse draws on the ideas proposed by political scientist Robert Putnam (1995), who famously showed that Americans in the 1990s went "bowling alone" rather than in the leagues that thrived during the mid-twentieth century.

5. Río de la Loza [1863] 1911: 212; Río de la Loza and Craveri 1858: 18; Orozco y Berra 1855: 91.

6. Río de la Loza [1863] 1911: 212.

7. Rivera Cambas 1880–1883, vol. 2: 330–31.

8. AHCM, Aguas, Excepción, Vol. 44, Exp. 1.

9. AHCM, Aguas, Mercedes en Arrendamiento, Vol. 71, Exp. 496; AHCM, Aguas, Mercedes en Arrendamiento, Vol. 70, Exp. 441; AHCM, Aguas, Mercedes en Arrendamiento, Vol. 71, Exp. 550; AHCM, Aguas, Mercedes en Arrendamiento, Vol. 75, Exp. 1073.

10. Scholarly works describing well drilling in France, Germany, and England circulated in the Mexican scientific community. Río de la Loza and Craveri 1858, citing Héricart de Thury 1829.

11. Talavera Ibarra 2004.

12. Peñafiel 1884: 48–49, 55; "Noticia Geológica" 1858; Barcena 1885: 262.

13. Kurlansky 2003: 27.

14. Río de la Loza and Craveri 1858: 17.

15. Barcena 1885: 266; Buenrostro 1875: 179.

16. Del Valle 1859: 25–26.

17. "Pozos Absorventes" 1856.

18. Peñafiel 1884; Talavera Ibarra 2004; Río de la Loza [1863] 1911: 220–21.

19. Talavera Ibarra 2004: 301; Peñafiel 1884: 153–54.

20. Talavera Ibarra 2004: 301.

21. Peñafiel 1884: 51; Barcena 1885.

22. Peñafiel 1884: 191.

23. Peñafiel 1884: 50.

24. Orvañanos 1895: 221.

25. Barcena 1884; Peñafiel 1884: 13–15.

26. Talavera Ibarra 2004: 296.

27. Orozco y Berra 1858; Peñafiel 1884: 50.

28. Peñafiel 1884: 12–14, 35–36.

29. AHCM, Aguas, Foráneas Chapultepec, Vol. 36, Exp. 153.

30. "Noticia Geológica" 1858: 27; Talavera Ibarra 2004.

31. Talavera Ibarra 2004: 303–5.

32. Río de la Loza [1863] 1911: 231.

33. AHCM, Aguas, Foráneas Chapultepec, Vol. 48, Exp. 22, "Sobre Propiedad de la Alberca de Chapultepec"; Río de la Loza and Craveri 1858: 89.

34. AHCM, Aguas, Foráneas Chapultepec, Vol. 48, Exp. 23, "Acuerdo de la Junta de Hacienda."

35. AHCM, Aguas, Foráneas Chapultepec, Vol. 48, Exp. 1 (June 11, 1820), "Visita de Ojos."

36. AHCM, Aguas, Foráneas Chapultepec, Vol. 48, Exp. 22 (September 30, 1870).

37. AHCM, Aguas, Foráneas Chapultepec, Vol. 48, Exp. 25, "Jose Amor y Escandón se queja"; AHCM, Aguas, Foráneas Chapultepec, Vol. 48, Exp. 29, "Varios vecinos del barrio de San Miguel Chapultepec."

38. Gray 1878; Haven 1875: 224–25.

39. AHCM, Aguas, Foráneas Chapultepec, Vol. 48, Exp. 10 (November 18, 1870).

40. Kingsley 1874: 260; Bárcenas 1885: 267.

41. Geiger 1874: 219–20.

42. Forbes 1851: 140.

43. Forbes 1851: 204.

44. Lyon 1828: 94.

45. Lyon 1828: 63.

46. Donnavan 1848: 22.

47. Domenech 1858: 272.

48. Gilliam 1846.

49. Forbes 1851: 139.

50. Geiger 1874: 48.

51. Ruxton 1847: 59.

52. Forbes 1851: 139.

53. Gilliam 1846: 202.

54. Conkling 1883: 212.

55. Blake and Sullivan 1888: 116.

56. Smith 1889: 64.

57. Crawford 1899: 61–63.

58. Crawford 1899: 62, italics original.

59. Jackson 1890: 210.

60. Baker 1895: 127.

61. Smith 1889: 64.

62. Blake and Sullivan 1888: 115–16.

63. Donnavan 1847: 22.

64. Case 1917: 85.

65. García Cubas 1904: 373.

66. Blake and Sullivan 1888: 40–42.

67. Ober 1883: 564–65.

68. Carson 1909: 136–37.

69. McCarty 1888: 214.

70. Beaufoy 1828: 70.

71. McCarty 1888: 183.

72. Geiger 1874; George 1877; Hale and Hale 1893.

73. Elton 1867: 30.

74. Iglehart 1887: 230.

75. Elton 1867: 30.

76. Elton 1867: 35.

77. Rivera Cambas 1880–1883, vol. 2: 286.

78. Prantl and Groso 1901: 39.

79. Rogers 1893: 143.

80. Rivera Cambas 1880–1883, vol. 2: 285.

81. Macías González 2012.

82. Website of the Sociedad de Autores y Compositores de México, http://sacm.org.mx/biografias/biografias-interior.asp?txtSocio = 03416 (retrieved August 12, 2016).

83. Rivera Cambas 1880: 284.

84. Paz 1882.

85. De Cuéllar 1889; Paz 1882.

86. Talavera Ibarra 2004.

87. AHCM, Mercedes en Arrendamiento, Vol. 143, Exp. 7000 (May 3, 1894); AHCM, Mercedes en Arrendamiento, Vol. 143, Exp. 7000 (September 3, 1894).

88. Williams 1992; Wiltse 2007.

89. AHCM, Policía en General, Vol. 3638, Exp. 991 (March 13, 1891).

90. AHCM, Policía en General, Vol. 3639, Exp. 1064 (1895).

91. AHCM, Policía en General, Vol. 3638, Exp. 991 (March 21, 1891).

92. AHCM, Policía en General, Vol. 3638, Exp. 991 (June 6, 1898).

93. Iglehart 1887: 275.

94. Rivera Cambas 1880–1883: 284.

95. Rivera Cambas 1880–1883: 284.

96. Prantl and Groso 1901: 39.

97. Peñafiel 1884, 1900.

98. Riva Palacio [1890] 2010.

99. Tenorio-Trillo 1996.

100. De Valle-Arizpe 1949: 412.

101. De Valle-Arizpe 1949: 425.

102. García Cubas 1904; Macías-González 2012.

103. George 1877: 128; Haven 1875: 224–25.

104. Pineda Mendoza 2000.

105. Rivera Cambas 1880–1883, vol. 2: 319; Gray 1878: 59.

106. García Cubas 1904: 373.

107. Gray 1878: 59.

108. AHCM, Aguas, Vol. 48, Exp. 37 (June 6, 1892), J. Ceballos to Ayuntamiento.

109. Crawford 1889: 179.

110. Mintz 1985.

111. "La 'Colonia de la Condesa'" (July 8, 1906), *El Mundo Ilustrado* 13(12).

112. Agostoni 2003, 2005.

6   CHEMISTRY, BIOLOGY, AND THE HETEROGENEITY OF MODERN WATERS

1. Aboites 2012; Aréchiga Córdoba 2004; Banister and Widdifield 2014; Linton 2010.

2. Latour 1993; Carrillo 2001.

3. Anderson and Tabb 2002; Green 1986; Mackaman 1998.

4. "Noticia Geológica" 1858.

5. Barcena 1885.

6. "Noticia Geológica" 1858.

7. In 1857, 200 wells in Mexico City produced 867 m3/hr, or 4.33 m3/hr per well. In 1883 there were 483 wells; at the same rate of 4.33 m3/hr, they produced 2,091 m3/hr. Total aqueduct flow at that time was 1,237 m3/hr (Peñafiel 1884: 50). By comparison, in 2013 the extraction of water from the aquifer of the Metropolitan Zone of Mexico City (Zona Metropolitana de la Ciudad de Mexico—ZMCM) was about 34 m3/second, or 122,400 m3/hr.

Extraction from the entire Valley in 2013 was approximately 58 m3/second, or 208,800 m3/hr, about twice the rate of recharge (Gómez Reyes 2013: 19).

8. Coley 1979, 1982; Porter 1990.

9. Miller 1962.

10. Pauer 1872.

11. Masson 1864.

12. Hay et al. 1870.

13. Urbán Martinez and Aceves Pastrana 2001: 37.

14. Río de la Loza and Craveri 1858: 17–18.

15. "Noticia Geológica" 1858: 28.

16. Río de la Loza and Craveri 1858: 11.

17. Río de la Loza [1863] 1911: 216–27.

18. Corbin 1988.

19. Río de la Loza [1863] 1911: 224.

20. Río de la Loza [1863] 1911: 225.

21. Barcena 1885.

22. Ross 2009: 575.

23. Peñafiel 1884: 121.

24. Peñafiel 1884: 122.

25. Mackaman 1998.

26. Hardy 1829: 503.

27. Hardy 1829: 416.

28. Hardy 1829: 416.

29. Gilliam 1846: 194.

30. Weiss and Kemble 1967: 5.

31. Durie 2006.

32. Moreno 1849; Sáez de Heredia 1849.

33. Sáez de Heredia 1849: 19.

34. Sáez de Heredia 1849; González Urueña 1849: 4.

35. Sáez de Heredia 1849: i.

36. Sáez de Heredia 1849: ii.

37. For example, Dumont and Torres 1762.

38. Sáez de Heredia 1849: 4.

39. Nogueras 1849: prologue.

40. González Urueña 1849.

41. González Urueña 1849: 17–24; Moreno 1849.

42. Lugo 1875: 7.

43. Lopez 2012; Schifter Aceves 2014.

44. "Valladolid" 1790: 207.

45. Aceves Pastrana 1993: 90.

46. De la Cal y Bracho 1832: ix.

47. Masson 1864: 209.

48. Masson 1864: 213.

49. Antomarchi 1835.

50. Gayuca 1843.

51. Xavier A. Lopez y de la Peña, "El último médico de Napoleon Bonaparte en Aguascalientes, México," http://medicinaaguascalientes.blogspot.com/search?updated-min = 2013–01–01T00:00:00–08:00&updated-max = 2014–01–01T00:00:00–08:00&max-results = 11 (retrieved July 7, 2016).

52. Masson 1864: 177.

53. Masson 1864: 175.

54. Tort 1858.

55. León 1882.

56. Jennings 2006; Rodríguez Sánchez 2006.

57. Lobato 1884.

58. Peñafield 1884.

59. Lobato 1884: xii.

60. Lobato 1884: 23.

61. Lobato 1884: xiii.

62. Lobato 1884: 188.

63. Lobato 1884: 98–99.

64. Lobato 1884: 193–94.

65. Lobato 1884: 213.

66. Lobato 1884: 212.

67. Lobato 1884: 94.

68. Rivera Cambas 1880–1883, vol. 2: 306.

69. *Gaceta Médica de México* 22 (1887): 70.

70. Delacroix 1865. I thank Eric Jennings for pointing this out.

71. "Acta Numero 6," *Gaceta Médica* 22 (1886): 38.

72. *Gaceta Médica* 22 (1886): 35–40, 47–55, 70.

73. Sosa 1889a.

74. Orvañanos 1895; Liceaga 1890.

75. Macías-González 2012: 29.

76. Jennings 2006.

77. Sosa 1889b: n.p.

78. *Anales del Instituto Médico Nacional* 5 (1894): 82.

79. "Forjadores" 2011; Escotto Velazaquez 1999; Liceaga 1949.

80. Liceaga and Gayol 1898: 842.

81. Liceaga 1949: 152; Archivo Histórico de la Secretaría de Salud (AHSS), Fondo Beneficencia Pública, Hospitales, Hospital General, Caja 2, Exp. 22 (1903), "Pago de útiles."

82. Lobato 1884: 100–106; Bárcenas 1885.

83. Liceaga 1892.

84. *Gaceta Médica* 26 (1891): 231–32; Ross 2009.

85. *Gaceta Médica* 26 (1891): 232.

86. Tenorio-Trillo 1996.

87. Liceaga 1892.

88. Prantl and Groso 1901: 40.

89. Liceaga 1892.

90. "El Balneario del Riego," *El Mundo Ilustrado* 1906; Armendaris 1902.

91. Sosa 1889b: n.p.

92. Aguilar 1885: 190.

93. García Cubas 1904: 372.

94. Lugo 1875; Huerta 1883; Michaus 1893; Marquez Landa 1901; Mayer 1906.

95. Mayer 1906: 11.

96. Castillejos 1908.

97. Thomas de la Peña 1999.

98. Prantl and Groso 1901: 40–41.

99. Mayer 1906: 3.

100. AHSS, Fondo Salubridad Publica (FSP), Sección Servicio Jurídico (SSJ), Caja 3, Exp. 11 (1924), "Proyecto de Reglamento de Baños Públicos del DF"; AHSS, FSP, SSJ, Caja 3, Exp. 16, Newspaper clipping: "El Baño más Higienico," source and date unknown.

101. Liceaga 1949: 84–88, 92–93.

102. AHSS, FSP, SSJ, Caja 3, Exp. 16, Newspaper clipping: "El Baño más Higienico," source and date unknown.

103. AHSS, FSP, SSJ, Caja 21, Exp. 9 (April 9, 1930), Transcript of Meeting.

104. AHSS, FSP, SSJ, Caja 3, Exp. 11 (1924), "Proyecto de Reglamento de Baños Públicos del DF."

105. AHSS, FSP, SSJ, Caja 3, Exp. 16 (December 10, 1924), "Memo."

106. AHSS, FSP, SSJ, Caja 3, Exp. 16.

107. AHSS, FSP, SSJ, Caja 3, Exp. 16 (January 23, 1926), Alberca Pane to Salubridad Pública.

108. "Tabla Analitica" 1858: 53.

109. Orozco y Berra 1855: 86.

7 DISPOSSESSION AND BOTTLING AFTER THE REVOLUTION

1. Marx [1867] 1990, ch. 26; Harvey 2005, ch. 4.

2. Arreguín Mañon 1998; Walsh 2008; Wolfe, 2017.

3. Mazzarella (2003) argues that the specific, physical qualities of commodities—the unique objects themselves, their use-values—are necessary to satisfy wants and thus are reinforced rather than erased by the otherwise homogenizing force of commodity exchange.

4. Lobato 1884: 187.

5. Hayes 2004.

6. "Las Aguas Minerales del Peñon," *Boletín de la Asociacion Financiera Internacional, International Edition, English-Spanish* 3, no. 1 (August 1907): 5.

7. Armendaris 1902; Bringas Nostti 2010.

8. "El Balneario del Riego," *El Mundo Illustrado* 13, no. 2 (1906): 427.

9. Bringas Nostti 2010.

10. Archivo Histórico de la Secretaría de Salud (AHSS), Fondo Salubridad Publica (FSP), Sección Servicio Jurídico (SSJ), Caja 8, Exp. 19 (1927), "Prohibición de comercializar las aguas gaseosas de Tehuacán"; AHSS, FSP, SSJ, Caja 11, Exp. 1 (August 5, 1927), "Medellín to Dr. Federico Falcón."

11. AHSS, FSP, SSJ, Caja 8, Exp. 19 (August 12, 1927), "Hijos de Montt o Salubridad Pub DF"; AHSS, FSP, SSJ, Caja 8, Exp. 19 (August 13, 1927), "Laboratorio General, Sección de Bacteriología"; AHSS, FSP, SSJ, Caja 8, Exp. 19 (October 10, 1927), "Federico Falcón to Jefe

Salubridad Publica DF"; AHSS, FSP, SSJ, Caja 8, Exp. 19 (October 18, 1927), "Secretario General to Señores Hijos de W. Montt."

12. AHSS, FSP, SSJ, Caja 11, Exp. 1 (September 1, 1927), "De la Garza to Jefe Salud Publica."

13. AHSS, FSP, SSJ, Caja 10, Exp. 2 (August 6, 1927), "Arturo Vargas to Jefe Consejo Superior de Salubridad."

14. AHSS, FSP, SSJ, Caja 11, Exp. 1 (November 23, 1927), "Agente Rafael Carrasco to Salubridad Pública."

15. AHSS, FSP, SSJ, Caja 9, Exp. 1 (August 2, 1927), "Luis Portillo to Jefe Salubridad Pública."

16. AHSS, FSP, SSJ, Caja 9, Exp. 1 (July 21, 1927), "Luis Portillo to Salubridad Publica"; AHSS, FSP, SSJ, Caja 9, Exp. 1 (August 31, 1927): "Luis Portillo to Salubridad Publica."

17. AHSS, FSJ, SSP, Caja 36, Exp. 24 (April 4, 1933), Ramón Ramirez, "Las Fabricas Clandestinas," pp. 1, 8.

18. AHSS, FSJ, SSP, Caja 36, Exp. 24, "Las Autoridades Deben Evitar el Contrabando y la Fabricación de Sacarina"; *El Universal* (March 28, 1933), pp. 1–2.

19. AHSS, FSP, SSJ, Caja 10, Exp. 6 (1927–30), "Dictámenes Servicio Jurídico . . . Servicio de Comestibles y Bebidas"; AHSS, FSP, SSJ, Caja 8, Exp. 19 (October 19, 1927), "Secretario General de Salubridad Publica to Delegado Federal de Puebla"; AHSS, FSP, SSJ, Caja 8, Exp. 19 (November 4, 1927), "Falcón to Salubridad Pública DF"; AHSS, FSP, SSJ, Caja 8, Exp. 19 (November 23, 1927), "Falcón to Jefe del Departmento de Salubridad Pública, DF"; AHSS, FSP, SSJ, Caja 9, Exp. 1 (July 21, 1927), "Luis Portillo to Salubridad Pública."

20. AHSS, FSJ, SSP, Caja 41, Exp. 15 (July 15, 1935), "Informe labores . . . sección cuarta."

21. AHSS, FSJ, SSP, Caja 36, Exp. 24 (1933), "Exp. Relacionado al Campaña de Prensa."

22. Arturo Mundet, "El Pro de la Salubridad," *El Universal* (March 24, 1933); "Las Fábricas Clandestinas," *El Universal* (April 4, 1933), pp. 1, 8.

23. "Nino Murió por Tomar un Refresco con Sacarina," *El Universal*, pp. 1, 5; AHSS, FSJ, SSP, Caja 36, Exp. 24 (April 5, 1933).

24. AHSS, FSJ, SSP, Caja 45, Exp. 2 (July 25, 1935), "Informe"; AHSS, FSJ, SSP, Caja 45, Exp. 2 (April 4, 1935), "Informe/Carta SP."

25. Chittenden 1884; Baur 1959.

26. Valenza 2000: 34–43.

27. Jones 1967.

28. Ballou 1890; Bates 1887; Blake and Sullivan 1888; Ford 1893; Jackson 1890; Margati 1885; Smith 1889.

29. Ober 1883: 565, 625.

30. Archivo Histórico del Agua, México (AHA), Aguas Superficiales (AS), Caja 4581, Exp. 60978; AHA, AS, Caja 4359, Exp. 57850 (March 1932), Servicio Consular Mexicano, Oficina de Presidio Texas, "Informe Comercial."

31. Garza Cantú 1892.

32. Randle received a credit from bankers in New York. AHA, Aguas Nacionales (AN) Caja 1195, Exp. 16636 (August 9, 1900), Contract, Comunidad SBTC and Slayden, www.tramz.com/mx/mo/mo.html (retrieved March 26, 2012).

33. Daniell 1892.

34. www.topochico.com/quien2.html (retrieved February 17, 2012).

35. Daniell 1892.

36. Hodson 1888: 776.

37. *Texas Health Journal* 4, no. 8 (February 1891): 235.

38. Cited in "Mineral Wells of Texas," *Texas Health Journal* 5, no. 12 (June 1893): 315; Valenza 2000: 41.

39. Mexican National Railroad Company 1893: 13.

40. Hodson 1888: 775.

41. AHA, AS, Caja 1195, Exp. 16636 (August 20, 1929); AHA, AS, Caja 1196, Exp. 16641 (April 12, 1930), Garza to Secretaría de Agricultura y Fomento (SAF), www.topochico.com/quien2.html (retrieved February 17, 2012). Randle was also the owner of Monterrey's tramway system (Mora-Torres 2001: 123).

42. Morris 1902: 46.

43. www.topochico.com/lideraz.html (retrieved February 17, 2012).

44. AHA, AN, Caja 469, Exp. 4947 (April 8, 1931), Carlota Zambrano to SAF; AHA, AN, Caja 469, Exp. 4947 (May 21, 1931), Alfonso de la Torre to SAF; AHA, AS, Caja 4905, Exp. 68434 (October 12, 1903), Pedro Treviño to SAF; AHA, AN, Caja 519, Exp. 5715 (August 28, 1903), "Informe."

45. Holden 1994.

46. AHA, AN, Caja 4581, Exp. 60978.

47. AHA, AS, Caja 4519, Exp. 69882 (May 29, 1911), SAF to Secretaría de Obras Publicas (SCOP).

48. AHA, AS, Caja 4905, Exp. 68434 (March 13, 1909), SCOP to SAF.

49. AHA, AS, Caja 1655, Exp. 24274 (December 28, 1926).

50. AHA, AS, Caja 271, Exp. 6547 (August 1, 1930), Celso Cepeda to SAF.

51. Glennon 2004; Wolfe 2013.

52. I thank Luis Aboites for this clarification.

53. AHA, AS, Caja 271, Exp. 6547 (August 1, 1930), Celso Cepeda to SAF; AHA, AS, Caja 1195, Exp. 16636 (October 23, 1929), "Solicitud de Dotacion de Derechos de Agua Caliente y Ojo Caliente."

54. AHA, AN, Caja 522, Exp. 5735 (July 24, 1930), "Informe."

55. AHA, AS, Caja 271, Exp. 6547 (August 1, 1930), Celso Cepeda to SAF.

56. AHA, AS; Caja 271, Exp. 6547 (August 1, 1930), Celso Cepeda to SAF.

57. AHA, AS, Caja 1665, Exp. 24274 (February 19, 1924), Comisión Nacional Agraria (CNA) to Celso Cepeda; AHA, AS, Caja 271, Exp. 6547 (August 1, 1930), Cepeda to SAF; AHA, AS, Caja 271, Exp. 6547 (February 25, 1924), CNA to Celso Cepeda.

58. AHA, AN, Caja 522, Exp. 5735 (September 7, 1926), Resolución Presidencial, Plutarco Elias Calles.

59. AHA, AN, Caja 522, Exp. 5735 (April 9, 1929), "Informe 438."

60. AHA, AS, Caja 1665, Exp. 24274 (December 28, 1926), Comisión Particular Administrativa Congregacion San Bernabe Topo Chico to CNA; AHA, AS, Caja 1665, Exp. 24274 (March 25, 1927), "Peticion."

61. AHA, AS, Caja 1665; Exp. 24274 (May 21, 1927); AHA, AS, Caja 1665, Exp. 24274 (June 7, 1927); AHA, AS, Caja 633, Exp. 9139 (Novmeber 30, 1927), "Informe 480."

62. AHA, AN, Caja 522, Exp. 5735 (December 5, 1927), SAF "Gestion #55."

63. AHA, AN, Caja 470, Exp. 4963 (December 31, 1927), SAF.

64. AHA, AS, Caja 271, Exp. 6547 (March 15, 1928).

65. AHA, AN, Caja 522, Exp. 5735 (July 24, 1930), "Informe" Ing Leonel Lemus.

66. AHA, AS, Caja 1195, Exp. 16636 (November 18, 1930), "Informe 575."

67. AHA, AS, Caja 1665, Exp. 24274 (February 25, 1929), Governor Jose Benitez to SAF.

68. AHA, AS, Caja 1665, Exp. 24272 (February 2, 1929), SAF to Garza Gonzalez.

69. AHA, AS, Caja 1196, Exp. 16641 (May 3, 1928), Contract Junta and Garza.

70. AHA, AS, Caja 1195, Exp. 16636 (October 23, 1929), Solicitud Derechos Agua Caliente y Ojo Caliente.

71. AHA, AN, Caja 522, Exp. 5735 (February 18, 1929), SAF Informe #75.

72. AHA, AN, Caja 522, Exp. 5735 (July 16, 1929), SAF Informe #368.

73. AHA, AS, Caja 1655, Exp. 24274 (December 28, 1926); AHA, AN, Caja 522, Exp. 5737 (June 20, 1929), Vidaurri to Parres.

74. AHA, AS, Caja 1665, Exp. 24274 (January 2, 1929), Garza to SAF.

75. AHA, AN, Caja 522, Exp. 5735 (July 16, 1929), SAF Informe #368.

76. AHA, AN, Caja 522, Exp. 5735 (September 7, 1929).

77. AHA, AS, Caja 1195, Exp. 16636 (August 20, 1929), Solicitud.

78. AHA, AS, Caja 1196, Exp. 16641 (September 21, 1929), Solicitud.

79. AHA, AS, Caja 1195, Exp. 16636 (September 23, 1929), SAF to Congregation.

80. AHA, AN, Caja 489, Exp. 5201 (May 31, 1930), SAF to Compañía.

81. AHA, AN, Caja 489, Exp. 5201 (October 10, 1929), Informe 485.

82. AHA, AN, Caja 469, Exp. 4947 (May 21, 1931), Alfonso de la Torre to SAF, www.topochico.com/lideraz.html (retrieved February 17, 2012).

83. www.topochico.com/lideraz.html (retrieved February 17, 2012).

84. AHA, AS, Caja 1196, Exp. 16641 (October 26, 1929), Acta Notarial; AHA, AN, Caja 522, Exp. 5735 (July 24, 1930), "Informe" Ing Leonel Lemus.

85. AHA, AS, Caja 271, Exp. 6547 (October 23, 1929), Consejo Superior de Salubridad to Congregation.

86. AHA, AS, Caja 271, Exp. 6547 (January 9, 1930), Cepeda to SAF; AHA, AS, Caja 11096, Exp. 16641 (October 10, 1930), CNA to SAF.

87. AHA, AN, Caja 522, Exp. 5735 (July 24, 1930), "Informe 278"; AHA, AS, Caja 11096, Exp. 16641 (October 10, 1930), CNA to SAF.

88. AHA, AN, Caja 522, Exp. 5735 (October 10, 1931), Cepeda to Cedillo.

89. www.topochico.com/lideraz.html (retrieved April 20, 2016).

90. Arca Continental Company, *Siempre Hacia Adelante, Informe Annual*, 2015.

91. www.topochicousa.net/#the-legend (retrieved September 27, 2016).

8   SPA TOURISM IN TWENTIETH-CENTURY MEXICO

1. Berger and Wood 2010.

2. Berger 2006.

3. Alonso Alvarez 2012; Anderson and Tabb 2002; Chambers 2002; Mackaman 1998; Walton 2012.

4. AHA, AS, Caja 723, Exp. 10513 (March 1, 1921), Tomas Lamadrid to SAF.

5. *Periódico Oficial*, Distrito Norte de Baja California 33, no. 8 (February 10, 1921); AHA, AS, Caja 723, Exp. 10513 (March 1, 1921), Tomas Lamadrid to SAF.

6. Gómez Estrada 2002.

7. Bringas Nostti 2010: 432–40.

8. Berger 2006.

9. Sackett 2010.

10. Arizmendi Domínguez 1999: 89, 101–3; AHA, AN, Caja 1204, Exp. 16337 (September 23, 1932).

11. AHA, AS, Caja 2058, Exp. 31075, pp. 187–96 (June 2, 1941), "Informe #79"; AGN, Presidentes, Ávila Camacho, 491/2 (May 22, 1941), Ortiz Rubio to Ávila Camacho.

12. Arizmendi Domínguez 1999: 79–93.

13. AHA, AN, Caja 836, Exp. 10373 (September 30, 1953), Union de Defensa Economico Social de Ixtapan de la Sal, México to Secretaría de Recursos Hidráulicos, Direccion de Aprovechamientos Hidráulicos, Concesiones y Vedas, Region Sur, México DF.

14. AHA, AN, Caja 836, Exp. 10373 (May 20, 1930), "Acta Numero 10"; AHA, AN, Caja 580, Exp. 6463, "Asunto Contestando Oficio"; Aboites and Estrada 2004: 243.

15. AHA, AS, Caja 2058, Exp. 31075, pp. 151–55, Vergara to Olivier Ortiz; AHA, AN, Caja 2058, Exp. 31075, "Testimonio," Juan Hernandez and Onofre Morales.

16. AHA, AN, Caja 1206, Exp. 16354, pp. 132–34 (August 14, 1943), CMT (Olivier Ortiz) to Secretario of Ag y Fom (Marte R Gómez).

17. AHA, AN, Caja 1206, Exp. 16354, pp. 178–95 (October 27, 1943), Transcript of the Junta de Avenencia.

18. *Gaceta del Gobierno del Estado de México* 40(33), May 18, 1932.

19. AHA, AS, Caja 2058, Exp. 31075, pp. 124–30 (March 15, 1939), "Memorándum sobre Valides del Contrato."

20. AHA, AN, Caja 2058, Exp. 31075 (1939), Escrituras de Propiedad (1933); AHA, AN, Caja 2058, Exp. 31075 (October 24, 1938), Solicitud de Confirmación de Derechos al Manantial Laguna Verde.

21. AHA, AN, Caja 2058, Exp. 31075 (November 2, 1938), Antonino del Pilar to SAF; AHA, AN, Caja 2058, Exp. 31075 (February 20, 1939), Graciano Sánchez to SAF.

22. AHA, AS, Caja 576, Exp. 6460, p. 108 (May 16, 1941), "Vergara to Enrique Flores Magón." For more on Magón, see Hart 1978.

23. AGN, Presidentes, Ávila Camacho, 491/2 (November 26, 1942), "Informe del Consejo presentado a la Asamblea General de Accionistas"; AHA, AN, Caja 1206, Exp. 16354, pp. 132–34 (August 14, 1943), "Compañía Mexicana de Turismo to Secretaría de Agricultura y Fomento."

24. AHA, AN, Caja 2058, Exp. 31075 (May 18, 1932); *Gaceta del Gobierno del Estado de México* 33, no. 40 (April 23, 1932).

25. See Berger (2006) for a discussion of Ortiz Rubio and his relation to Luis Montes de Oca, Banco de México president and promoter of tourism who organized credit for building hotels.

26. AGN, Presidentes, Ávila Camacho, 491/2 (February 21, 1941), "Ortiz Rubio to Ávila Camacho."

27. AHA, AS, Caja 2058, Exp. 31075, pp. 187–96 (June 2, 1941), "Informe #79."

28. AGN, Presidentes, Ávila Camacho, 491/2 (May 22, 1941), "Ortiz Rubio to Ávila Camacho"; AHA, AS, Caja 2058, Exp. 31075, pp. 259–60 (March 6, 1941), "CMT to Ávila Camacho."

29. AHA, AN, Caja 2058, Exp. 31075 (June 13, 1941).

30. AGN, Presidentes, Ávila Camacho, 491/2 (July 28, 1941), Ávila Camacho to Ortiz Rubio.

31. AHA, AS, Caja 2058, Exp. 31075, pp. 273–76 (April 29, 1942), "Informe #74."

32. AHA, AN, Caja 580, Exp. 6463, pp. 109–10 (December 31, 1941), Memorandum, Jefe de Departamento de Aguas.

33. AHA, AN, Caja 580, Exp. 6463, pp. 138–41 (April 27, 1942), Flores Magón to Dirección General de Aguas, SAF.

34. AHA, AN, Caja 580, Exp. 6463, p. 143 (June 15, 1942), Javier Gaxiola to Sociedad Cooperativa.

35. AHA, AN, Caja 1206, Exp. 16354 (May 15, 1942), Solicitud de Concesión, Manantial Santa Catarina; AHA, AN, Caja 1206, Exp. 16354, p. 4 (July 16, 1942), Carlos Rodríguez to SAF.

36. AHA, AN, Caja 1206, Exp. 16354, pp. 119–23 (July 7, 1943), C. Rodríguez to Director Dept Aguas SAF.

37. AHA, AN, Caja 1206, Exp. 16354, p. 8, Municipal President Ixtapan de la Sal to Carlos Rodríguez; AHA, AN, Caja 1206, Exp. 16354, p. 107 (October 30, 1942), Carlos Rodríguez to SAF.

38. AGN, Presidentes, Ávila Camacho, 491/2 (April 9, 1943), "Memorandum," Ortiz Rubio to Ávila Camacho; AGN, Presidentes, Ávila Camacho, 491/2 (May 23, 1943), Marte R. Gómez (SAF) to Ávila Camacho; AHA, AN, Caja 1206, Exp. 16354, pp. 132–34 (August 14, 1943), Olivier Ortiz to SAF.

39. AHA, AN, Caja 1206, Exp. 16354 (January 18, 1944), Telegram, Ayuntamiento to SAF; AHA, AN, Caja 1206, Exp. 16354 (January 17, 1944), Dario Hernandez to Ávila Camacho.

40. AHA, AN, Caja 1206, Exp. 16354 (July 3, 1944).

41. AHA, AN, Caja 1206, Exp. 16354, pp. 213–30 (February 15, 1944), "Informe #40."

42. Daniela Barragán (January 22, 2015), "Perfil: ¿Quien es la familia San Román, beneficiada en Edomex y afín a los Peña?" *sinembargo.mx*, www.sinembargo.mx/22-01-2015/1225941 (retrieved September 21, 2016).

43. AGN, Presidentes, Ávila Camacho, 491/2 (August 18, 1944), Ortiz Rubio to Ávila Camacho.

44. AGN, Presidentes, Ávila Camacho, 491/2 (August 18, 1944), Ortiz Rubio to Ávila Camacho.

45. AHA, AN, Caja 1203, Exp. 16337 (September 14, 1945), "Informe #442."

46. AHA, AN, Caja 1203, Exp. 16337 (May 3, 1946).

47. AHA, AN, Caja 1203, Exp. 16337 (August 23, 1945), "Informe"; AHA, AN, Caja 1203, Exp. 16337 (September 14, 1945), "Informe #442"; AHA, AN; Caja 1203; Exp. 16337 (February 7, 1946), Dirección General de Aguas to SAF; AHA, AN, Caja 1204, Exp. 16337 (June 7, 1946), Governor of Mexico State to Arturo San Román.

48. AHA, AN, Caja 1206, Exp. 16354 (August 22, 1945).

49. AHA, AN, Caja 1206, Exp. 16354 (October 16, 1946), Rodríguez to Dirección General de Aguas.

50. AHA, AN, Caja 1206, Exp. 16354 (October 16, 1948), Rodríguez to SRH; AHA, AN, Caja 1206, Exp. 16354 (March 9, 1949), SRH to Rodríguez; AHA, AN, Caja 1206, Exp. 16354 (March 15, 1949), Rodríguez to SRH; AHA, AN, Caja 1206, Exp. 16354 (April 2, 1951), Rodríguez to SRH; AHA, AN, Caja 1206, Exp. 16354 (April 24, 1951), Dirección General de Aguas to Carlos Rodríguez.

51. AHA, AN, Caja 1206, Exp. 16354 (November 27, 1953), Informe.

52. Cruz Jimenez et al. 2012.

53. AHA, AN, Caja 1204, Exp. 16337 (August 15, 1960), Poblado Yerbasbuenas to SRH; AHA, AN, Caja 1203, Exp. 16337 (August 27, 1967), Isauro Lugo to Comisión rio Balsas.

9 VIRTUOUS WATERS IN THE TWENTY-FIRST CENTURY

1. Gordillo 2014.

2. Interview, April 6, 2013, www.spamexico.com/history.php.

3. www.cantarranas.com.mx/spaserv.html, website of the Spa of the Hotel Cantarranas, Tehuacán (retrieved October 25, 2016).

4. www.tehuacanusa.com/index-1.html, website of the Tehuacán bottled water company (retrieved October 25, 2016).

5. Gleick 2010.

BIBLIOGRAPHY

ARCHIVES CONSULTED

AFIIE. Archivo Fotográfico del Instituto de Investigaciones Estéticas, Universidad Nacional Autónoma de México, Mexico City, Mexico.

AGN. Archivo General de la Nación, Mexico City, Mexico.

AHA. Archivo Histórico del Agua, Mexico City, Mexico.

AHCM. Archivo Histórico de la Ciudad de México, Mexico City, Mexico.

AHENL. Archivo Histórico del Estado de Nuevo León, Monterrey, Nuevo León, Mexico.

AHENM. Archivo Histórico de la Escuela Nacional de Medicina, Mexico City, Mexico.

AHSS. Archivo Histórico de la Secretaría de Salud, Mexico City, Mexico.

AMIS. Archivo Municipal, Ixtapan de la Sal, Estado de México, Mexico.

AMT. Archivo Municipal, Tehuacán, Puebla, Mexico.

BHNM. Biblioteca y Hemeroteca Nacional de México, Universidad Nacional Autónoma de México, Mexico City, Mexico.

HMNFR. Hemeroteca Nacional de México, Mexico City, Mexico.

PERIODICALS

American Wine Press and Mineral Water News, USA

Anales del Instituto Médico, Nacional, Mexico

Boletín de la Asociacion Financiera Internacional, International Edition, English-Spanish, Mexico

El Estudio, Mexico

El Mundo Ilustrado, Mexico

El Universal, Mexico

Gaceta del Gobierno del Estado de Mexico, Mexico

Gaceta Médica, Mexico

Gazeta de México, Mexico
Periódico Oficial, Distrito Norte de Baja California, Mexico
Revista Mexicana Periódico Científico y Literario, Mexico
Suplemento al Tomo Sesto del Boletín de la Sociedad Mexicana de Geografía y Estadística, Mexico
The Sanitarian, USA
The Texas Health Journal, USA

BIBLIOGRAPHY

Aboites, L. 1987. *La irrigación revolucionaria: Historia del sistema nacional de riego del río Conchos, Chihuahua, 1927–1938.* Mexico: SEP.
——. 1998. *El agua de la nación: Una historia política de México, 1888–1946.* Mexico: CIESAS.
——. 2012. "The Transnational Dimensions of Mexican Irrigation, 1900–1950." *Journal of Political Ecology* 19: 70–80.
—— and Valeria Estrada. 2004. *Del agua municipal al agua nacional: Materiales para una historia de los municipios en México, 1901–1945.* Mexico: CIESAS / Archivo Histórico del Agua / Comisión Nacional del Agua / El Colegio de México.
Aceves Pastrana, Patricia. 1993. *Química, botánica y farmacia en la Nueva España a finales del siglo XVIII.* Mexico: UAM Xochimilco.
——. 1996. "Tradición y modernidad en la Nueva España: Estudios sobre aguas minerales (S. XVII–XVIII)." *LLULL* 19: 325–45.
Adams, Robert McCormick. 1966. *The Evolution of Urban Society.* Chicago: Aldine.
Agostoni, Claudia. 2003. *Monuments of Progress: Modernization and Public Health in Mexico City, 1876–1910.* Calgary, Boulder, Mexico: University of Calgary Press, University Press of Colorado, UNAM.
——. 2005. "Las delicias de la limpieza: La higiene en la Ciudad de México." In *Historia de la Vida Cotidiana en México, Tomo IV, Bienes y Vivencias, El Siglo XIX,* edited by Anne Staples, 563–97. México: Fondo de Cultural Económica.
"Aguas Minerales de Peñón" [advertisement]. 1907 (August). *Boletín de la Asociación Financiera Internacional* 3, no. 1. International Edition, English/Spanish.
Aguilar, Federico. 1885. *Último Año de Residencia en México.* Bogotá: Imprenta de Ignacio Borda.
Aguilera Klink, Federico. 1998. "Economía y cultura del agua: Algunas reflexiones." *Demófilo, Revista de Cultura Tradicional de Andalucía* 27: 65–83.
——. 1999. "Hacia una Nueva Economía del Agua: Cuestiones Fundamentales." In *El agua a debate desde la universidad: Hacia una nueva cultura del agua,* edited by Pedro Arrojo and F. Javier Martínez-Gil, 49–66. Zaragoza, Spain: Navarro and Navarro.
Alberro, Solange. 1992. Del Gachupín al criollo: O de como los españoles de México dejaron de serlo. Mexico: El Colegio de México.
Alcina Franch, José. 1994. "Plantas medicinales para el temascal mexicano." *Estudios de Cultura Náhuatl* 24. Mexico: UNAM.
——, Andres Ciudad Ruiz, and Joséfa Iglesias Ponce de León. 1980. "El 'temazcal' en Mesoamérica: Evolución, forma y función." *Revista Española de Antropología Americana* 10: 93–132.

Alonso Alvarez, Luis. 2012. "The Value of Water: The Origins and Expansion of Thermal Tourism in Spain, 1750-2010." *Journal of Tourism History* 4(1): 15-34.

Alzate Ramírez, José Antonio. [1792] 1894. *Gacetas de Literatura de México, Tomo 2.* Mexico: Oficina Tipográfica de la Secretaría de Fomento.

"American Public Health Association." 1892. *The Sanitarian* 29(277): 485.

Anderson, James. 2002. *Daily Life during the Spanish Inquisition.* Westport, CT: Greenwood.

Anderson, Susan, and Bruce Tabb. 2002. *Water, Leisure and Culture: European Historical Perspectives.* Oxford and New York: Berg.

Antomarchi, Francois Carlo. 1835. "Aguas Sulfurosas y Gaseosas de Xochitepec." *Revista Mexicana Periódico Científico y Literario* 1(1): 376-77, 505-7. Mexico: Ignacio Cumplido.

Apenes, Ola. 1944. "The Primitive Salt Production of Lake Texcoco, Mexico." *Ethnos* 1: 35-39.

Archibald, Elizabeth. 2012. "Bathing, Beauty and Christianity in the Middle Ages." *Insights* 5(1). Durham University, Institute of Advanced Study.

Aréchiga Córdoba, Ernesto. 2004. "De la exuberancia al agotamiento: Xochimilco y el agua, 1882-2004." In *A la orilla del agua: Política, urbanización y medio ambiente—Historia de Xochimilco en el siglo XX,* edited by María Eugenia Terrones, 95-149. México: Gobierno del Distrito Federal / Instituto Mora.

———. 2007. "Educación, propaganda o 'dictadura sanitaria': Estrategias discursivas de higiene y salubridad públicas en el México posrevolucionario, 1917-1945." *Estudios de Historia Moderna y Contemporánea de México* 33: 57-88.

Arizmendi Domínguez, Luis René. 1999. *Ixtapan de la Sal: Monografía Municipal.* Toluca: Instituto Mexiquense de Cultura.

Armendaris, Dr. Eduardo. 1902. "Estudio sobre las aguas de Tehuacán (Estado de Puebla)." Mexico: Secretaría de Fomento, Colonización e Industria, Instituto Médico Nacional.

Arreguín Mañon, José. 1998. *Aportes a la historia de la geohidrología en México, 1890-1995.* Mexico, DF: CIESAS.

Arrojo, Pedro. 2006. *El reto ético de la nueva cultura del agua: Funciones, valores y derechos en juego.* Barcelona: Ediciones Paidós Ibérica.

Aveleyra Arroyo de Anda, Luis. 2005. *El Peñón de los Baños y la leyenda de Copil.* Mexico: INAH.

"Aviso importante que da D. Andrés Caballero a las personas que necesitan ocurrir a los Baños del Peñol." *Gazeta de México,* Part 1: 6, no. 78 (November 19, 1794), 651-52; Part 2: 6, no. 79 (November 19, 1794), 654-55.

Bacon, William. 1997. "The Rise of the German and the Demise of the English Spa Industry: A Critical Analysis of Business Success and Failure." *Leisure Studies* 16(3): 173-87.

Bailey, Anne, and Josép Llobera, eds. 1981. *The Asiatic Mode of Production: Science and Politics.* London: Routledge & Kegan Paul.

Baker, Frank Collins. 1895. *A Naturalist in Mexico.* Chicago: David Oliphant.

Ballou, M. 1890. *Aztec Land.* Boston and New York: Houghton Mifflin.

Bando de Revillagigedo. 1793. *Gazeta de México,* August 31, pp. 428-32.

Banister, Jeffrey, and Stacie Widdifield. 2014. "The Debut of 'Modern Water' in Early 20th-Century Mexico City: The Xochimilco Potable Waterworks." *Journal of Historical Geography* 46: 36-52.

Barcena, Maríano. 1885. *Tratado de geología.* Mexico: Oficina Tipográfica de la Secretaría de Fomento.

Barkin, David, and Timothy King. 1970. *Regional Economic Development: The River Basin Approach in Mexico.* Cambridge: Cambridge University Press.

Bates, J. H. 1887. *Notes of a Tour in Mexico and California.* New York: Burr Printing House.

Baur, John. 1959. *The Health-Seekers of Southern California, 1870–1900.* Pasadena, CA: Huntington Library.

Beaufoy, Mark. 1828. *Mexican Illustrations Founded upon Facts; Indicative of the Present Condition of Society, Manners, Religion, and Morals among the Spanish and Native Inhabitants of Mexico.* London: Carpenter and Son.

Beaumont, Pablo de la Purísima Concepción. 1772. *Tratado de la agua mineral caliente de San Bartholome.* Mexico: Imprenta Joséph Antonio de Hogal.

Bennett, Jane. 2010. *Vibrant Matter: A Political Ecology of Things.* Durham, NC: Duke University Press.

Berger, Dina. 2006. *The Development of Mexico's Tourism Industry: Pyramids by Day, Martinis by Night.* New York: Palgrave/Macmillan.

——— and Andrew Grant Wood, eds. 2010. *Holiday in Mexico: Critical Reflections on Tourism and Tourist Encounters.* Durham, NC: Duke University Press.

Birrichaga Gardida, Diana. 2004. "El dominio de las 'aguas ocultas y descubiertas': Hidráulica colonial en el centro de México, siglos XVI–XVII." In *Mestizajes tecnológicos y cambios culturales en México,* edited by Enrique Florescano and Virginia García Acosta. México: CIESAS/Porrúa.

Blake, Mary Elizabeth, and Margaret Frances Sullivan. 1888. *Mexico: Picturesque, Political, Progressive.* Boston and New York: Lee and Shepard.

Blakeslee, Donald. 2010. *Holy Ground, Healing Water: Cultural Landscapes at Waconda Lake, Kansas.* College Station: Texas A&M University Press.

Blázquez, J. M., and M. P. García-Gelabert. 1992. "Recientes aportaciones al culto de las aguas en la Hispania Romana." *Espacio, Tiempo y Forma, Serie 2, Historia Antigua* 5: 21–66.

Bonfil, Guillermo. 1987. *México profundo: Una civilización negada.* México, DF: SEP.

Boyer, C., ed. 2012. *A Land between Waters: Environmental Histories of Modern Mexico.* Tucson: University of Arizona Press.

Brewer, Harry. 2004. "Historical Perspectives on Health: Early Arabic Medicine." *Journal of the Royal Society for the Promotion of Health* 124(4): 184–87.

Bringas Nostti, Raúl. 2010. *Historia de Tehuacán: De yiempos prehispánicos a la modernidad.* Mexico: Miguel Angel Porrúa.

Brockliss, L. W. B. 1990. "The Development of the Spa in Seventeenth-Century France." *Medical History* 34(S10): 23–47.

Brooks, G. F. 1891 (February). "Dr. G. F. Brooks." *Texas Health Journal* 4(8): 235.

Buenrostro, Felipe. 1875. *Historia del segundo congreso constitucional de la República Mexicana que funcionó en los años de 1861, 62 y 63, Tomo 2.* Mexico: Imprenta Poliglota.

Caballero, Andrés. 1794. "Aviso importante que da D Andrés Caballero a las Personas que necesitan ocurrir a los Baños del Peñol." *Gazeta de México* 6, no. 78 (November 19, 1794), 651–52; 6, no. 79 (November 19, 1794), 653–55.

Calderón de la Barca, Mrs Frances Erskine Inglis. 1843. *Life in Mexico during a Residence of Two Years in that Country,* vol. 1. Boston: Charles Little and James Brown.

Candiani, Vera. 2012. "The Desague Reconsidered: Environmental Dimensions of Class Conflict in Colonial Mexico." *Hispanic American Historical Review* 91(1): 5–39.

———. 2014. *Dreaming of Dry Land: Environmental Transformation in Colonial Mexico City.* Stanford, CA: Stanford University Press.

Cárdenas, Juan de. [1591] 1945. *Colección de incunables americanos Siglo XVI volumen IX: Problemas y secretos maravillosos de las Indias.* Madrid: Ediciones Cultura Hispanica.

Carreón Nieto, María del Carmen. 1999. *Las expediciones científicas en la Intendencia de Valladolid.* Morelia: Universidad Michoacana de San Nicolás de Hidalgo.

Carrillo, Ana María. 2001. "Los comienzos de la bacteriología en México." *Elementos* 42: 23–27.

Carson, William E. 1909. *Mexico: The Wonderland of the South.* Detroit: MacMillan.

"Carta de Don Gabriel de Ocampo." *Gazeta de México* 6, no. 79 (November 19, 1794), 656–60.

Casal García, Raquel, and Silvia González Soutelo. 2010. *Os balnearios de Galicia: Orixe e desnevolvemento.* Santiago de Compostela: Universidade de Santiago de Compostela.

Case, Alsden. 1917. *Thirty Years with the Mexicans: In Peace and Revolution.* New York: Fleming H. Revell.

Castillejos, Manuel J. 1908. "Algunas consideraciones terapéuticas sobre los baños." Thesis, National School of Medicine, Mexico.

Cátedra Tomás, María. 2009. "El agua que cura." *Revista de Dialectología y Tradiciones Populares* 64(1): 177–210.

Cayleff, Susan. 1987. *Wash and Be Healed: The Water-Cure Movement and Women's Health.* Philadelphia: Temple University Press.

Cervantes de Salazar, Francisco. 1985. *Crónica de la Nueva España.* Mexico: Porrúa.

Chambers, Thomas. 2002. *Drinking the Waters: Creating an American Leisure Class at Nineteenth-Century Mineral Springs.* Washington and London: Smithsonian Institution Press.

Childe, V. Gordon. 1950. "The Urban Revolution." *Town Planning Review* 21(1): 3–17.

Chittenden, Newton H. 1884. *Health Seekers', Tourists' and Sportsmen's Guide to the Sea-side, Lake-side, Mountain and Mineral Spring Health and Pleasure Resorts of the Pacific Coast.* San Francisco: C.A. Murdock.

Clark, Scott. 1994. *Japan: A View from the Bath.* Honolulu: University of Hawaii Press.

Coley, Noel. 1979. "'Cures without Care': 'Chymical Physicians' and Mineral Waters in Seventeenth-Century English Medicine." *Medical History* 23: 191–214.

———. 1982. "Physicians and the Chemical Analysis of Mineral Waters in Eighteenth-Century England." *Medical History* 26: 123–44.

Comaroff, Jean, and John Comaroff. 2001. *Millennial Capitalism and the Culture of Neoliberalism.* Durham, NC: Duke University Press.

Conkling, Howard. 1883. *Mexico and the Mexicans.* New York: Tainter Brothers, Merrill.

Consejo Nacional de Turismo. 1966. *Los Balnearios Medicinales de México.* Mexico: Editorial Libros de México.

Cook, Scott. 2004. *Understanding Commodity Cultures: Explorations in Economic Anthropology with Case Studies from Mexico.* Lanham, MD: Rowman & Littlefield.

Corak, Sandra, and Irena Ateljevic. 2007. "Colonisation and 'Taking the Waters' in the 19th Century: The Patronage of Royalty in Health Resorts of Opatija, Habsburg Empire and

Rotorua, New Zealand." In *Royal Tourism: Excursions around Monarchy*, edited by Philip Long and Nicola J. Palmer. Clevedon, Buffalo, Toronto: Channel View.

Corbin, Alain. 1988. *The Foul and the Fragrant: Odor and the French Social Imagination*. Translated by Miriam L. Kochan, Roy Porter, and Christopher Predergast. Cambridge, MA: Harvard University Press.

———. 1994. *The Lure of the Sea: The Discovery of the Seaside in the Western World*. Cambridge: Cambridge University Press.

Crawford, Cora Hayward. 1889. *The Land of the Montezumas*, second edition. New York: John B. Alden.

Crehan, Kate. 2002. *Gramsci, Culture and Anthropology*. Berkeley: University of California Press.

———. 2016. *Gramsci's Common Sense: Inequality and Its Narratives*. Durham, NC: Duke University Press.

Cresson, Frank M. Jr. 1938. "Maya and Mexican Sweat Houses." *American Anthropologist* 40(1): 88–104.

Cruz Jiménez, Graciela, Cecilia Cadena Insotroza, and Rocío del Carmen Serrano Barquín. 2012. "La transición de una comunidad agrícola a turística: Ixtapan de la Sal, México." *Revista Rosa dos Ventos* 4(11): 222–34.

Cumplido, Ignacio. 1838. "Baños." *Calendario Tercero Portátil*. Mexico: I. Cumplido.

———. 1842. "Baños Públicos." *Séptimo Calendario de Cumplido*. México: I. Cumplido.

Daniell, Lewis E. 1892. "Jules A. Randle, Monterey, Mexico." In *Personnel of the Texas State Government with Sketches of Representative Men of Texas*, 527–29. San Antonio: Maverick Printing House.

Dávalos, Marcela. 1997. *Basura e ilustración: La limpieza de la Ciudad de México a fines del siglo XVIII*. Mexico: INAH/Departamento del Distrito Federal.

De Cuéllar, José. 1889. "Baile y cochino: Novela de costumbres Mexicanas." In *La Linterna Mágica: Segunda Época, Tomo 1*, tercera edición. Barcelona: Tipo-litografía de Espasa y Compañía.

De Gortari Rabiela, H., 2012. "La ciudad de México de finales del siglo XVIII: Un diagnóstico desde la 'ciencia de la policía.'" *Historia Contemporánea* 24: 115–35.

Delacroix, Emile. 1865. *Les Eaux: Etude Hygiénique et Médicale*. Paris: Chez F. Savy.

Deleuze, Gilles, and Felix Guattari. 1987. *A Thousand Plateaus: Capitalism and Schizophrenia*. Minneapolis: University of Minnesota Press.

"Departamento de Baños." 1906. *El Mundo Ilustrado* 13, vol. 2, no. 12.

De la Cal y Bracho, Antonio. 1832. *Ensayo para la Materia Médica Mexicana*. Puebla: Oficina del Hospital de S. Pedro.

De Valle-Arizpe, Artemio. 1949. *Calle vieja y calle nueva*. México: Editorial Jus.

Del Valle, Juan N. 1859. *El Viajero en México: O sea la Capital de la Republica Encerrada en un Libro*. Mexico: Tipografía de M. Castro.

Díaz, Plácido. 1876. "Estudio Sobre la Phthisis, y Acción que en ella Ejercen las Aguas Thermo-Minerales de Puebla." *El Estudio* 1: 7–11, 79–82, 109–12, 175–77, 372–75.

Díaz del Castillo, Bernal. [1568] 1844. *The Memoirs of the Conquistador Bernal Díaz del Castillo, Vol. 1 (of 2) Written by Himself Containing a True and Full Account of the Discovery and Conquest of Mexico and New Spain*. Translated by John Ingram Lockhart.

London: J Hatchard and Son. Project Gutenberg, Ebook #32474, www.gutenberg.org/files/32474/32474-h/32474-h.htm#Footnote_64_65 (retrieved April 19, 2015).

Domenech, Emmanuel. 1858. *Missionary Adventures in Texas and Mexico: A Personal Narrative of Six Years' Sojourn in those Regions*. London: Longman, Brown, Green, Longmans and Roberts.

Donnavan, Corydon. 1848. *Adventures in Mexico; Experienced during a Captivity of Seven Months*, twelfth edition. Boston: George R. Holbrook.

Dumont, Joséph, and Nicolás de Torres. 1762. *Virtudes de las aguas del Peñol*. Mexico: Imprenta de la Biblioteca Mexicana [Archivo Histórico de la Escuela Nacional de Medicina, Mexico].

Durie, Alistair. 2006. *Water Is Best: The Hydros and Health Tourism in Scotland, 1840–1940*. Edinburgh: John Donald.

"Edificios de "El Peñón." 1906. *El Mundo Ilustrado* 13, vol. 2, no. 12.

"El Balneario del Riego." 1906. *El Mundo Ilustrado* 13, vol. 2, no. 12.

"El Peñón, Las Aguas mas Saludables del Pais." 1906 (September 16). *El Mundo Ilustrado* 13, vol. 2, no. 12.

Elton, James Frederick. 1867. *With the French in Mexico*. London: Chapman and Hall.

Endfield, Georgina. 2012. "The Resilience and Adaptive Capacity of Social-Environmental Systems in Colonial Mexico." *PNAS* 109(10): 3676–81.

Erfurt-Cooper, Patricia, and Malcolm Cooper. 2009. *Health and Wellness Tourism: Spas and Hot Springs*. Bristol, Buffalo, and Toronto: Channel View.

Escobar, Arturo. 1999. "After Nature: Steps to an Antiessentialist Political Ecology." *Current Anthropology* 40(1): 1–30.

Escotto Velazquez, Jorge. 1999. "Semblanza del Doctor Eduardo Liceaga." *Revista Médica del Hospital General* 62(4): 237–39.

Ewald, Ursula. 1985. *The Mexican Salt Industry: 1560–1980*. Stuttgart and New York: Gustav Fischer Verlag.

Foley, Ronan. 2010. *Healing Waters: Therapeutic Landscapes in Historic and Contemporary Ireland*. Farnham, UK, and Burlington, VT: Ashgate.

Forbes, Alexander. 1851. *A Trip to Mexico, or Recollections of a Ten-Months Ramble in 1849–1850*. London: Smith, Elder.

Ford, I. 1893. *Tropical America*. New York: Charles Scribner's Sons.

"Forjadores de la Ciencia en México—Eduardo Liceaga." 2011. Quiminet website, October 18. www.quiminet.com/articulos/forjadores-de-la-ciencia-en-mexico-eduardo-liceaga-2601278.htm (retrieved July 3, 2014).

Foucault, Michel. 1991. "Governmentality." In *The Foucault Effect: Studies in Governmentality*, edited by Graham Burchell, Colin Gordon, and Peter Miller. Chicago: University of Chicago Press.

———. 2009. *Security, Territory, Population: Lectures at the College de France, 1977*. Edited by Michel Sennellart. New York: Picador.

Gandy, Matthew. 2002. *Concrete and Clay: Reworking Nature in New York City*. Cambridge, MA, and London: MIT Press.

———. 2014. *The Fabric of Space: Water, Modernity, and the Urban Imagination*. Cambridge, MA: MIT Press.

García Cubas, Antonio. 1904. *El libro de mis recuerdos: Narraciones históricas, anecdóticas y de costumbres mexicanas anteriores al actual estado social*. México: Imprenta de Arturo García Cubas.

García Samper, Asunción. 1997. "La industria de la sal y de la cerámica en la región de Coxcatlan, Tehuacán, vista por las fuentes etnohistóricas y arqueológicas, Siglos XII al XVI." In *Simposium internacional Tehuacán y su entorno: Balance y perspectivas*, edited by Eréndira de la Lama. Mexico: INAH.

García Sánchez, Magdalena. 2004. "El modo de vida lacustre en el valle de México: ¿Mestizaje o proceso de aculturación?" In *Mestizajes tecnológicos y cambios culturales en México*, edited by Enrique Florescano and Virginia García Acosta. Mexico: CIESAS/ Porrúa.

Garza Cantú, Rafael. 1892. "Algo sobre la infección palustre en Monterrey." In *Estudio sobre la higiene en Monterrey*, edited by Rafael Garza Cantú, 1–20. Monterrey: printed by the author.

Gayuca, Regino. 1843. "Simple análisis de las aguas de Xochitepec." In *El museo Mexicano, o miscelánea pintoresca de amenidades curiosas e instructivas*, 336. Mexico DF: Ignacio Cumplido.

Geels, Frank. 2005. "Co-evolution of Technology and Society: The Transition in Water Supply and Personal Hygiene in the Netherlands (1850–1930)—A Case Study in Multi-level Perspective." *Technology in Society* 27: 363–97.

Geertz, Clifford. 1973. "The Wet and the Dry: Traditional Irrigation in Bali and Morocco." *Human Ecology* 1(1): 23–39.

Geiger, John Lewis Geiger. 1874. *A Peep at Mexico: Narrative of a Journey across the Republic from the Pacific to the Gulf in December 1873 and January 1874*. London: Trubner.

Gelles, Paul. 1994. "Channels of Power, Fields of Contention: The Politics of Irrigation and Land Recovery in an Andean Peasant Community." In *Irrigation at High Altitudes: The Social Organization of Water Control Systems in the Andes*. Society for Latin American Anthropology Publication Series 12, edited by William Mitchell and David Guillet, 233–74. Washington, DC: American Anthropological Association.

——. 2000. *Water and Power in Highland Peru: The Cultural Politics of Water and Development*. New Brunswick, NJ: Rutgers University Press.

George, Alfred. 1877. *Holidays at Home and Abroad*. London: W. J. Johnson.

Gibson, Charles. 1964. *The Aztecs under Spanish Rule: A History of the Indians of the Valley of Mexico, 1519–1810*. Stanford, CA: Stanford University Press.

Gil de Arriba, Carmen. 2000 (August 1). "La difusión social y espacial del modelo balneario: De la innovación medica al desarrollo de las practicas de ocio." *Scripta Nova* 69(40). www.ub.es/geocrit/sn-69-40.htm.

Gilliam, Albert. 1846. *Travels over the Table Lands and Cordilleras of Mexico during the Years 1843 and '44*. Philadelphia: John Moore.

Gillpatrick, Wallace. 1912. *The Man Who Likes Mexico*. New York: Century.

Gilly, A. 1971. *La revolución interrumpida*. México: El Caballito.

Girón Irueste, Fernando. 2006. "Uso Médico del agua en el mundo Hispánico bajo medieval (siglos XII–XV)." *Balnea* 1: 79–95.

Gleick, P. H. 2003. "Global Freshwater Resources: Soft-Path Solutions for the 21st Century." *Science* 302(5650): 1524–28.

Gleick, Peter. 2010. *Bottled and Sold: The Story behind Our Obsession with Bottled Water.* Washington, DC: Island Press.

Glennon, Robert. 2004. *Water Follies: Groundwater Pumping and the Fate of America's Fresh Waters.* Washington, DC: Island Press.

Gómez Estrada, José Alfredo. 2002. *Gobierno y casinos: El origen de la riqueza de Abelardo L Rodríguez.* Mexicali: UABC/Instituto Mora.

Gómez Reyes, Eugenio. 2013. "Valoración de las componentes del balance hídrico usando información estadística y geográfica: La cuenca del Valle de México." *Realidad, Datos y Espacio: Revista Internacional de Estadística y Geografía* 4(3): 4–27.

González Jacome, Alba. 2007. *Agua y Agricultura: Ángel Palerm, la Discusión con Karl Wittfogel sobre el modo asiático de producción y la construcción de un modelo para el estudio de Mesoamerica.* Mexico: Universidad Iberoamericana.

González Uruena, Juan Manuel. 1849. *La hidropatía o más bien, la hidroterapia, desde su origen hasta nosotros examinada bajos sus dos aspectos vulgar y científico.* Mexico: Tipografía de R. Rafael.

Goubert, Jean-Pierre. 1989. *The Conquest of Water: The Advent of Health in the Industrial Age.* Princeton, NJ: Princeton University Press.

Gramsci, Antonio. 1971. *Selections from the Prison Notebooks of Antonio Gramsci.* Edited and translated by Quintin Hoare and Geoffrey Nowell Smith. New York: International.

Gran Balneario del Peñón, México. 1937. *Baños termominerales como tratamiento de algunas enfermedades.* Mexico: Gran Balneario del Peñón.

Gray, Albert Zabriskie. 1878. *Mexico as It Is: Being Notes of a Recent Tour in that Country.* New York: E. P. Dutton.

Green, Harry. 1986. *Fit for America: Health, Fitness, Sport and American Society.* New York: Pantheon Books.

Greenberg, James, and Thomas Park. 1994. "Political Ecology." *Journal of Political Ecology* 1(1): 1–12.

Groark, Kevin. 1997. "To Warm the Blood, to Warm the Flesh: The Role of the Steambath in Highland Maya (Tzeltal-Tzotzil) Ethnomedicine." *Journal of Latin American Lore* 20(1): 3–96.

Gruzinski, Serge. 1985. "Las cenizas del deseo: Homosexuales novohispanos a mediados del siglo XVII." In *De la santidad a la perversión: O de porque no se cumplía la ley de Dios en la sociedad novohispana,* edited by Sergio Ortega, 255–81. Mexico: Grijalbo.

Guillet, David W. 1992. *Covering Ground: Communal Water Management and the State in the Peruvian Highlands.* Ann Arbor: University of Michigan Press.

Hale, Edward, and Susan Hale. 1893. *A Family Flight through Mexico.* Boston: D. Lothrop.

Hamlin, Christopher. 1990. "Chemistry, Medicine, and the Legitimization of English Spas, 1740–1840." *Medical History* 34(S10): 67–81.

———. 2000. "'Waters' or 'Water'?: Master Narratives in Water History and Their Implications for Contemporary Water Policy." *Water Policy* 2: 313–25.

Hardy, Robert William. 1829. *Travels in the Interior of Mexico in 1825, 1826, 1827, & 1828.* London: Henry Colburn and Richard Bentley.

Harley, David. 1990. "A Sword in a Madman's Hand: Professional Opposition to Popular Consumption in the Waters Literature of Southern England and the Midlands, 1570–1870." *Medical History* 34(S10): 48–55.

Hart, John Mason. 1978. *Anarchism and the Mexican Working Class.* Austin: University of Texas Press.

———. 1987. *Revolutionary Mexico: The Coming and Process of the Mexican Revolution.* Berkeley: University of California Press.

———. 2002. *Empire and Revolution: The Americans in Mexico since the Civil War.* Berkeley: University of California Press.

Harvey, David. 2005. *The New Imperialism.* Oxford: Oxford University Press.

Hauser-Schaublin, Brigitta. 2003. "The Precolonial Balinese State Reconsidered: A Critical Evaluation of Theory Construction on the Relationship between Irrigation, the State, and Ritual." *Current Anthropology* 44(2): 153–81.

Haven, Gilbert. 1875. *Our Next-Door Neighbor: A Winter in Mexico.* New York: Harper and Brothers.

Hay, Guillermo, Alfonso Herrero, Manuel Rio de la Loza, G. Mendoza, and Leopoldo Rio de la Loza. 1870. "Informe de la comisión sobre las aguas potables en México." In *La Naturaleza: Periódico Científico de la Sociedad Mexicana de Historia Natural,* vol. 1, 6–15. Mexico: Ignacio Escalante y Cia.

Hayes, Constance. 2004. *The Real Thing: Truth and Power at the Coca-Cola Company.* New York: Random House.

Henshaw, Henry. 1910. "Sweating and Sweat-Houses." *Handbook of Northamerican Indians North of Mexico, Part 2. Smithsonian Institution Bureau of American Ethnology Bulletin no. 30:* 660–62. Washington, DC: Government Printing Office.

Héricart de Thury, L. E. F. 1829. *Considerations geologiques et physiques sur la cause de jaillisemente des eaux des puits fores ou fontaines articieielles et recherches sur l'orgine ou l'invention de la sonde, l'etata de l'art du fontenier-soundeur, et le degree de probabilite du success des puits fores.* Paris: Bachelier.

———. 1830. "Observations on the Cause of the Spouting of Overflowing Wells or Artesian Fountains." *Edinburgh New Philosophical Journal* 9: 157–65.

Hernández Alcántara, Martín. 2005. "El 'agua de Tehuacán' ha dejado de ser un emblema de ese municipio poblano." *La Jornada de Oriente,* July 20. www.lajornadadeoriente.com.mx/2005/07/20/puebla/pue2.html (retrieved December 24, 2017).

Hodgson, Susan Fox. 2004. "A Beautiful Spa: Thermal Waters at San Bartolo Agua Caliente, Mexico." *Geotermia* 17(1) (July–September): 45–50.

Hodson, Charles. 1888. "In Northeastern Mexico." *Catholic World* 46(276) (March): 775–76.

Holden, R. 1994. *Mexico and the Survey of Public Lands.* Dekalb: University of Northern Illinois Press.

Houston, Stephen. 1996. "Symbolic Sweatbaths of the Maya: Architectural Meaning in the Cross Group at Palenque." *Latin American Antiquity* 7(2): 132–51.

Huerta, Ramón. 1883. "Algunas consideraciones sobre hidroterapia." Thesis, National School of Medicine, Mexico.

Hunt, Robert C. 1988. "Size and the Structure of Authority in Canal Irrigation Systems." *Journal of Anthropological Research* 44(4): 335–55.

———, et al. 1976. "Canal Irrigation and Local Social Organization [and Comments and Reply]." *Current Anthropology* 17(3): 389–411.

Iglehart, Fanny. 1887. *Face to Face with the Mexicans: The Domestic Life, Education, Social and Business Ways.* New York: Fords, Howard and Hulbert.

"Informe de los trabajos ejecutados en la Sección 4a del Instituto Médico Nacional, Noviembre 1893." 1894. *Anales del Instituto Médico Nacional, Continuacion de El Estudio* 1: 193–95. Mexico: Secretaría de Fomento.

Jackson, Julia Newell. 1890. *A Winter Holiday in Summer Lands.* Cuba: A.C. McClurg.

Jackson, Ralph. 1990. "Waters and Spas in the Classical World." *Medical History* 34(S10): 1–13.

Jarrasse, Dominique. 2002. "La Importancia del termalismo en el nacimiento y desarrollo del turismo en Europa en el Siglo XIX." *Historia Contemporánea* 25: 33–49.

Jennings, Eric. 2006. *Curing the Colonizers: Hydrotherapy, Climatology, and French Colonial Spas.* Durham, NC: Duke University Press.

Johanson, Jens Christian. 1997. "Holy Springs and Protestantism in Early Modern Denmark: A Medical Rationale for a Religious Practice." *Medical History* 41: 59–69.

Johnson, Lyman, and Sonya Lipsett-Rivera, eds. 1998. *The Faces of Honor: Sex, Shame and Violence in Colonial Latin America.* Albuquerque: University of New Mexico Press.

Jones, B. M. 1967. *Health-Seekers in the Southwest, 1817–1900.* Norman: University of Oklahoma Press.

Joséph, Gil, and Daniel Nugent, eds. 1994. *Everyday Forms of State Formation: Revolution and the Negotiation of Rule in Modern Mexico.* Durham, NC: Duke University Press.

Kaplan, Martha. 2008. "Fijian Water in Fiji and New York: Local Politics and a Global Commodity." *Cultural Anthropology* 22(4): 685–706.

Kauffman, K. M. 1959. *The Baths of Pozzuoli: A Study of the Medieval Illuminations of Peter of Eboli's Poem.* London: Faber and Faber.

Kingsley, R. G. 1874. *South by West: Or, Winter in the Rocky Mountains and Spring in Mexico.* London: W. Isbister.

Kohn, Eduardo. 2013. *How Forests Think: Toward an Anthropology beyond the Human.* Berkeley: University of California Press.

Kupel, D. E. 2006. *Fuel for Growth: Water and Arizona's Urban Environment.* Tucson: University of Arizona Press.

Kurlansky, Mark. 2003. *Salt: A World History.* New York: Penguin.

"La 'Colonia de la Condesa.'" 1906. *El Mundo Ilustrado* 13, vol. 2, no. 12.

Landa, Fray Diego de. [1566] 1973. *Relación de las cosas de Yucatan*, tenth edition. Mexico: Editorial Porrúa.

Lanning, John Tate. 1983. *The Royal Protomedicato: The Regulation of the Medical Profession in the Spanish Empire.* Durham, NC: Duke University Press.

Lansing, J. Stephen. 1987. "Balinese 'Water Temples' and the Management of Irrigation." *American Anthropologist* 89(2): 326–41.

Lara García, María Pepa. 1997. *La cultura del agua: Los baños públicos en Málaga.* Malaga: Editorial Sarria.

Latour, Bruno. 1993. *The Pasteurization of France.* Cambridge, MA: Harvard University Press.

———. 2005. *Reassembling the Social: An Introduction to Actor-Network Theory.* Oxford: Oxford University Press.

Lempa, Heikki. 2002. "The Spa: Emotional Economy and Social Classes in Nineteenth-Century Pyrmont." *Central European History* 35(1): 37–73.

León, Simón. 1882. "Etude sur les eaux du Peñón de los Baños." *Bibliothèque Homéopathique* 13(9): 367–78.

Le Beuf, Dr. L. G. 1905. "The New General Hospital of Mexico City." *New Orleans Medical and Surgical Journal* 57: 767–68.

Liceaga, Eduardo. 1890. *Le plateau central du Mexique (Mesa Central), considere comme station sanitaire pour les phtisiques*. Berlin: H. S. Hermann.

———. 1892. *Essay on the Peñón Waters and Baths: A Study Presented to the National Academy of Medicine*. Mexico City: Military Industrial School Printing Office.

———. 1949. *Mis recuerdos de otros tiempos*. Edited by Francisco Fernendez del Castillo. Mexico: Talleres Gráficos de la Nación.

——— and Roberto Gayol. 1898. "Proyecto de hospital general de la Ciudad de México." In *Memorias del 2 Congreso Médico Pan-americano verificado en la Ciudad de México, D.F., Noviembre 16,17,18,19 de 1896*. México: Hoeck y Compañía.

Linton, Jamie. 2010. *What Is Water?: The History of a Modern Abstraction*. Vancouver: UBC Press.

Lobato, José. 1884. *Estudio sobre los aguas medicinales de la República Mexicana*. México: Secretaría de Fomento.

Lopes Brenner, Eliane. 2005. "El desarrollo turístico de la región de aguas termales de Goiás, Brasil." *Cuadernos de turismo* 16: 105–21.

López Austin, Alfredo. 1985. *La educación de los antiguos Nahuas 1*. Mexico: SEP/Ediciones El Caballito.

———. 2012. *Cosmovision y pensamiento indigena*. Mexico: IIS, UNAM.

López de Gomara, Francisco. [1552] 1966. *Historia general de las Indias*. Barcelona: Editorial Iberia.

Lopez, Rick. 2012. "Nature as Subject and Citizen: The Royal Botanical Expedition to New Spain (1787–1803)." In *A Land between Waters: Environmental Histories of Modern Mexico*, edited by Chris Boyer. Tucson: University of Arizona Press.

Lozano y Castro, M. 1900. "Análisis del agua que producen los pozos de la Penitenciaria." *Anales del Instituto Médico Nacional* 4: 303.

Lugo, J. M. 1875. "Estudio clínico de la hidroterapia racional." Thesis, National School of Medicine, Mexico.

Lyon, G. F. 1828. *Journal of a Residence and Tour in the Republic of Mexico in the Year 1826: With Some Account of the Mines of that Country*, vol. 1. London: J. Murray.

Mabry, Jonathan B., ed. 1996. *Canals and Communities: Small-Scale Irrigation Systems*. Tucson: University of Arizona Press.

Macías-González, Victor Manuel. 2012. "The Bathhouse and Male Homosexuality in Porfirian Mexico." In *Masculinity and Sexuality in Modern Mexico*, edited by Victor Manuel Macías González and Anne Rubenstein. Albuquerque: University of New Mexico Press.

Mackaman, Douglas. 1998. *Leisure Settings: Bourgeois Culture, Medicine, and the Spa in Modern France*. Chicago: University of Chicago Press.

Maldonado Polo, José Luis. 2000. "La expedición botánica a Nueva España, 1786–1803: El jardín botánico y la catedra de botánica." *Historia Mexicana* 50(1) (July–September): 5–56.

Mansen, Elisabeth. 1998. "An Image of Paradise: Swedish Spas in the Eighteenth Century." *Eighteenth-Century Studies* 31(4): 511–16.

Margadant, Guillermo Floris. 1986. "La Ordenanza de Intendentes para la Nueva España, ilusiones y logros." In *Memoria del IV Congreso de Historia del Derecho Mexicano, Tomo II*, edited by Beatriz Bernal. Mexico: UNAM.

Margati, J. 1885. *A Trip to the City of Mexico*. Boston: Putnam, Messervy.

Marquez Landa, Manuel. 1901. "La hidroterapia en las fiebres eruptivas." Thesis, National School of Medicine, Mexico.

Martinez Gil, Francisco Javier. 1996. *La nueva cultura del agua en España*. Bilbao: Editorial Bakeaz.

Martínez, Tomás, and Jacinta Palerm Viqueira, eds. 2000. *Antología sobre pequeño riego, Volumen II: Organizaciones autogestivas*. México: Colegio de Posgraduados/Plaza y Valdes.

Marx, Karl. [1846] 1998. *The German Ideology*. Amherst, NY: Prometheus Books.

———. [1867] 1990. *Capital, Volume 1*. London: Penguin Classics.

Mason, John Alden. 1935. "Mexican and Mayan Sweat Baths." *University of Pennsylvania Museum Bulletin* 6(2): 65–69.

Masson, Ernesto. 1864. *Olla podrida condimentada en México*. Paris: Imprenta Hispano-americana de Cosson y Compañía.

Matarredona Desantes, Nuria. 2014. "La arquitectura del baño de vapor en la cultura Maya." *Estudios de Cultura Maya* 44: 11–40.

Mayer, Oscar. 1906. "Nociones de hidroterapia científica." Thesis, National School of Medicine, Mexico.

Mazzarrella, William. 2003. *Shoveling Smoke: Advertising and Globalization in Contemporary India*. Durham, NC: Duke University Press.

McCarty, Joséph H. 1888. *Two Thousand Miles through the Heart of Mexico*. New York: Phillips and Hunt / Cincinnati: Cranston and Stove.

McCully, Patrick. 2001. *Silenced Rivers: The Ecology and Politics of Large Dams*, updated and revised edition. London: Zed Books.

Melosi, Martin. 1999. *The Sanitary City: Urban Infrastructure in America from Colonial Times to the Present*. Baltimore: Johns Hopkins University Press.

Melville, Roberto, and Scott Whiteford, eds. 2002 *Protecting a Sacred Gift: Water and Social Change in Mexico*. San Diego: Center for US-Mexican Studies.

Mendizabal, Miguel Othón de. 1925. "El Santuario de Chalma." *Anales del Museo Nacional de Arqueología, Historia y Etnografía, cuarta época* 3: 96–106.

Mexican National Railroad Company. 1893. *Tropical Tours to Toltec Towns: Presented with Compliments of the Mexican National Railroad. The Shortest, Quickest and Most Picturesque Route between Mexico and the United States*. New York: Mexican National Railroad.

Michaus, Salvador. 1893. "Algunas palabras acerca de la acción fisiológica terapéutica e higiénica de la hidroterapia." Thesis, National School of Medicine, Mexico.

Miller, Genevieve. 1962. "'Airs, Waters, and Places' in History." *Journal of the History of Medicine* 17(1): 129–40.

Mintz, Sidney. 1985. *Sweetness and Power: The Place of Sugar in Modern History*. New York: Penguin.

Mitchell, William P. 1976. "Irrigation and Community in the Central Peruvian Highlands." *American Anthropologist* 78(1): 25–44.

Mora, Gloria. 1992. "La literatura médica clásica y la arquitectura de las termas medicinales." *Espacio, Tiempo y Forma, Serie 2, Historia Antigua* 5: 121–32.

Morales Padrón, Francisco. 1949. "Baños termales en México (S. XVIII)." *Anuario de estudios americanos* 6: 695–713.

Moreno, José María 1849. *Del agua considerada como higiénica y medicinal, o de la hidrote-rapia por H. Seoutetten.* Translated and augmented by José Maria Moreno. Mexico: Manuel F Redondas.

Morris, [Ida Dorman]. "Mrs. J. E. Morris." 1902. *A tour in Mexico; illustrated from photographs taken en route by James Edwin Morris.* New York, London: Abbey Press.

Mosse, David. 1999. "Colonial and Contemporary Ideologies of 'Community Management': The Case of Tank Irrigation Development in South India." *Modern Asian Studies* 33(2): 303–38.

Munoz Lumbier, Manuel. 1934. *Las aguas medicinales en México.* Mexico, DF: Secretaría de Educación Pública.

Musset, Alain. 1992. *El agua en el Valle de México, Siglos XVI–XVIII.* Mexico: Centro de Estudios Mexicanos y Centroamericanos.

Nelson, Mike. 1998. *Spas and Hot Springs in Mexico,* second edition. Access Publishers Network.

Nogueras, José. 1849. *La medicina en el agua: O sea, la hidropatía que contiene.* Mexico: Imprenta de la Voz de la Religión.

Noriega Hernandez, Joana Cecilia. 2004. "El baño temascal Novohispano, de Moctezuma a Revillagigedo: Reflexiones sobre las practicas de higiene y expresiones de sociabilidad." Thesis, Universidad Autónoma Metropolitana, Ixtapalapa, Mexico.

"Noticia geológica del pozo abierto por los Sres. Pane y Molteni en los meses de Octubre y Noviembre de 1858 en la Calle de Sta. Catarina Num. 2 de esta Ciudad, con el sistema llamado Chino." 1858. *Suplemento al Tomo Sesto del Boletín de la Sociedad Mexicana de Geografía y Estadística,* 17–28. Mexico: Andres Boix.

Ober, Frederick A. 1883. *Travels in Mexico and Life among the Mexicans.* Boston: Estes and Lauriat.

Ocampo, Gabriel de. 1794. "Carta de Don Gabriel de Ocampo, doctor en medicina por esta Real y Pontifica Universidad, escrita a Don Andrés Caballero sobre las virtudes de los Baños del Penol." *Gazeta de México* 6, no. 78 (November 19): 656–60.

O'Dell, Tom. 2010. *Spas: The Cultural Economy of Hospitality: Magic and the Senses.* Lund: Nordic Academic Press.

Olivera, Ana. 1988. "Riesgo y salud en los Cuestionarios Americanos." In *Cuestionarios para la formación de las relaciones geográficas de las Indias,* edited by Francisco de Solano, 65–78. Madrid: CSIC.

Olmos, Ricardo. 1992. "Iconografía y culto a las aguas de época Prerromana en los mundos colonial e Ibérico." *Espacio, Tiempo y Forma, Serie 2, Historia Antigua* 5: 103–20.

Orozco y Berra, Manuel. 1855. *Apéndice al diccionario universal de historia y de geografía: Colección de artículos relativos a la República Mexicana . . . ,* vol. 1. Mexico: Imprenta de JM Andrade y F Escalante.

Orvañanos, Domingo. 1895. "Algunos datos sobre aguas públicas del Valle de México." *Gazeta Médica de México* 32: 219–21.

Ostrom, Elinor. 1992. *Crafting Institutions for Self-Governing Irrigation Systems.* San Francisco: ICS Press.

——— and Roy Gardner. 1993. "Coping with Asymmetries in the Commons: Self-Governing Irrigation Systems Can Work." *Journal of Economic Perspectives* 7(4): 93–112.

Palerm, Angel. 1952. "La civilización urbana." *Historia Mexicana* 2(2): 184–209.

———. 1955. "The Agricultural Bases of Urban Civilization in Mesoamerica." In *Irrigation Civilizations: A Comparative Study*, edited by Julian Steward et al. Washington, DC: PanAmerican Union.

———. 1973. *Obras hidráulicas Prehispánicas en el sistema lacustre del Valle de México.* Mexico, DF: Instituto Nacional de Antropología e Historia / SEP.

Palmer, Richard. 1990. "'In this our Lightye and Learned Tyme': Italian Baths in the Era of the Renaissance." *Medical History* 34(S10): 14–22.

Pane, Sebastián. 1856 (January 3). "Pozos absorbentes: Se permite haga investigaciones." In *Legislación Mejicana, o sea colección completa de las leyes, decretos y circulares que se han expedido desde la consumación de la Independencia, tomo que comprende de Enero a Junio de 1856*, 14–15. Mexico: Imprenta de Juan R. Navarro.

Pauer, L. 1872. *Aguas minerales fabricadas por L. Pauer en la Botica del Refugio (Bajos del Hotel Gual) con explicación de su empleo y sus efectos medicinales.* Mexico: Litografía Alemana.

Paz, Ireneo. 1882. *Nuevo guía del viajero en México para 1883.* Mexico: Ireneo Paz.

Peñafiel, Antonio. 1884. *Memoria sobre las aguas potables de la capital de México.* México: Oficina Tipográfica de la Secretaría de Fomento.

———. 1900. *Teotihuacán, estudio histórico y arqueológico.* Mexico: Secretaría de Fomento.

Pineda Mendoza, Raquel. 2000. *Origen, vida y muerte del Acueducto de Santa Fe.* Mexico: UNAM.

Pisani, Donald. 2002. *Water and American Government: The Reclamation Bureau, National Water Policy, and the West, 1902–1935.* Berkeley: University of California Press.

Pliny. 1949–54. *Natural History.* Translated by H. Rackham (vols. 1–5, 9), W. H. S. Jones (vols. 6–8), and D. E. Eichholz (vol. 10). Cambridge, MA: Harvard University Press.

Porter, Roy. 1990. Introduction to "The Medical History of Waters and Spas." *Medical History* 34(S10): vii–xii.

Prantl, Adolfo, and José L. Groso. 1901. *La Ciudad de México: Novísima guía universal de la capital de la Republica.* Mexico: J. Buxo.

Pulido Esteva, Diego. 2011. "Policía: Del buen gobierno a la seguridad, 1750–1850." *Historia Mexicana* 60(3): 1595–642.

Putnam, Robert. 1995. "Bowling Alone: America's Declining Social Capital." *Journal of Democracy* 6(1): 65–78.

Quintela, María Manuel. 2004. "Saberes e práticas termais: Uma perspectiva comparada em Portugal (Termas de S. Pedro de Sul) e no Brasil (Caldas da Imperatriz)." *Historia, Ciencias, Saude—Manguinhos* 11 (suplemento 1): 239–60.

Reisner, Marc. 1993. *Cadillac Desert.* New York: Penguin USA.

Rena, Antonio. 1966. *Aguas medicinales en los balnearios de México.* Mexico: Consejo Nacional de Turismo.

"Report of the Proceedings of the American Health Association, 1892." 1893. *Eighth Annual Report of the State Board of Health of the State of Kansas*, vol. 8, 253–60. Topeka: Hamilton Printing.

Riley, J. J. 1972. *A History of the American Soft Drink Industry: Bottled Carbonated Beverages, 1807–1957.* New York: Arno Press.

Rio de la Loza, Leopoldo. 1863. "Apuntes relativos a las fuentes brotantes o pozos artesianos." *Boletín de la Sociedad de Geografía y Estadística*, 1ª época, 10: 61–68. [In Rio de la Loza 1911: 222–31]

———. 1911. *Escritos de Leopoldo Rio de la Loza, compilados por el Señor Farmacéutico Juan Manuel Noriega, y publicados por la Secretaría de Instrucción Pública y Bellas Artes*. Mexico: Imprenta de Ignacio Escalante.

——— and Ernesto Craveri. 1858. "Opúsculo sobre los pozos artesianos y las aguas naturales de más uso en la Ciudad de México." In *Suplemento al Tomo Sesto del Boletín de la Sociedad Mexicana de Geografía y Estadística*, 9–17. Mexico: Imprenta de Andres Boix.

Riva Palacio, Vicente. [1890] 2010. *México a través de los siglos*. Mexico: Universidad Autónoma Metropolitana.

Rivera Cambas, Manuel. 1880–1883. *México pintoresco, artístico y monumental*. 3 vols. Mexico: Imprenta de la Reforma.

Rodríguez, Martha Eugenia. 2000. *Contaminación e insalubridad en la ciudad de México en el siglo XVIII*. Mexico: Universidad Nacional Autónoma de México.

Rodríguez Rivera, Virginia. 1945. "Baños Mágicos." *Revista Hispánica Moderna* 11(1–2): 170–80.

Rodríguez Sánchez, Juan Antonio. 2001. "Antecedentes históricos: la(s) memoria(s) del agua." In *Las aguas minerales en España*, edited by Juana Baeza Rodríguez-Caro, Juan Antonio López Geta, and Antonio Ramírez Ortega, 1–16. Madrid: Instituto Geológico y Minero de España.

———. 2006. "Institucionalización de la hidrología médica en España." *Balnea* 1: 25–40.

Rogers, Thomas. 1893. *México? Sí Señor*. Boston: Mexican Central Railway.

Rojas, Teresa. 1988. *Las siembras de ayer: La agricultura indígena del siglo XVI*. Mexico: SEP/CIESAS.

———. 1993. *La agricultura chinampera: Compilación histórica*, second edition. Mexico: Universidad de Chapingo.

Romero Contreras, Alejandro Tonatiuh. 2001. "Visiones sobre el temazcal mesoamericano: Un elemento cultural polifacético." *Ciencia Ergo Sum* 8(2): 133–34.

Romero-Huesca, Andrés, and Julio Ramírez Bollas. 2003. "La atención médica en el Hospital Real de Naturales." *Cirugía y Cirujanos* 71(6): 496–503.

Ross, Paul. 2009. "Mexico's Superior Health Council and the American Public Health Association: The Transnational Archive of Porfirian Public Health, 1887–1910." *Hispanic American Historical Review* 89(4): 573–602.

Ruiz Somavilla, María José. 1992. "Los valores sociales, religiosos y morales en las respuestas higiénicas de los siglos XVI y XVII: El problema de los baños." *DYNAMIS: Acta Hispanica ad Medicinea Scientarumque Historiam Illustrandam* 12: 155–87.

———. 1993. *El cuerpo limpio: Análisis de las prácticas higiénicas en la España del mundo moderno*. Málaga: University of Málaga, Service of Publications and Scientific Dissemination.

———. 2011. "El temazcal mesoamericano: Un modelo de adaptación cultural." *Nuevo Mundo Mundos Nuevos* [online], Cuestiones del tiempo presente, November 30, 2011, http://nuevomundo.revues.org/62198 (retrieved February 10, 2014).

Ruxton, George. 1847. *Adventures in Mexico and the Rocky Mountains*. London: John Murray.

Sackett, Andrew. 2010. "Fun in Acapulco? The Politics of Development on the Mexican Riviera." In *Holiday in Mexico: Critical Reflections on Tourism and Tourist Encounters*, edited by Dina Berger and Andrew Grant Wood. Durham, NC: Duke University Press.

Sáez de Heredia, Emeterio. 1849. *Hydropathia: O el uso medicinal del agua fría—Escrito según el sistema de Vicente Priessnitz.* Mexico: Tipografía de R. Rafael.

"Salón de embotellado de las Aguas." 1906. *El Mundo Ilustrado* 13, vol. 2, no. 12.

Sauder, Robert. 1994. *The Lost Frontier: Water Diversion in the Growth and Destruction of Owens Valley Agriculture.* Tucson: University of Arizona.

Schendel, Gordon. 1968. *Medicine in Mexico: From Aztec Herbs to Betatrons.* Austin: University of Texas Press.

Schifter Aceves, Liliana. 2014. "Las farmacopeas Mexicanas en la construcción de la identidad nacional." *Revista Mexicana de Ciencias Farmacéuticas* 45(2): 43–54.

Schwartzmental, John. 2015. *The Routledge Guidebook to Gramsci's Prison Notebooks.* New York: Routledge.

Seite, Sophie. 2013. "Thermal Waters as Cosmeceuticals: La Roche-Posay Thermal Spring Water Example." *Clinical, Cosmetic and Investigational Dermatology* 6: 23–28.

Seoutetten, H. 1849. *Del agua considerada como higiénica y medicinal, o de la hidroterapia.* Translated and augmented by José Maria Moreno. Mexico: Manuel F. Redondas.

Shadow, Robert, and María Rodríguez. 1990. "Símbolos que amarran, símbolos que dividen: Hegemonía e impugnación en una peregrinación campesina a Chalma." *Mesoamérica* 19: 33–72.

Silva Prada, Natalia. 2002. "El uso de los baños temascales en la visión de dos Médicos novohispanos: Estudio introductorio y transcripción documental de los informes de 1689." *Historia Mexicana* 52(1): 5–56.

Smith, Francis. 1889. *A White Umbrella in Mexico.* Boston and New York: Houghton Mifflin.

Smith, Michael E. 2009. "V. Gordon Childe and the Urban Revolution: A Historical Perspective on a Revolution in Urban Studies." *Town Planning Review* 80(1): 3–29.

Solano, Francisco de. 1988. *Cuestionarios para la formación de las relaciones geográficas de las Indias.* Madrid: CSIC.

Sosa, Secundino E. 1889a. "Nuestro programa." *El Estudio: Seminario de Ciencias Medicas* 1(1): 1–2.

———. 1889b. "Las aguas minerales de México." *El Estudio: Seminario de Ciencias Medicas* 1(30): 465–67.

Steward, Julian. 1949. "Cultural Causality and Law: A Trial Formulation of the Development of Early Civilizations." *American Anthropologist* 51(1): 1–27.

———, et al., eds. 1955. *Irrigation Civilizations: A Comparative Study.* Washington, DC: Pan American Union.

Strang, Veronica. 2004. *The Meaning of Water.* New York: Bloomsbury.

Swyngedouw, Eric, María Kaika, and Estaban Castro. 2002. "Urban Water: A Political-Ecology Perspective." *Built Environment* 28(2): 124–37.

"Tabla analítica de las aguas más usadas en la Ciudad de México." 1853. In *Suplemento al Tomo Sesto del Boletín de la Sociedad Mexicana de Geografía y Estadística*, 53. Mexico: Imprenta de Andrés Boix.

Talavera Ibarra, Oziel Ulises. 2004. "Los pozos artesianos en la Ciudad de México en la segunda mitad del Siglo XIX (1850–1880)." In *Miradas recurrentes I: La Ciudad de México en los siglos XIX y XX,* edited by María del Carmen Collado, 294–310. Mexico: Instituto Mora/UAM.

Tenorio-Trillo, Mauricio. 1996. *Mexico at the World's Fairs: Crafting a Modern Nation.* Berkeley: University of California Press.

Theimer-Sachse, Ursula. 2000. "Sobre higiene y medicina entre los Zapotecas durante la época de la conquista Española." *Indiana* 16: 185–210.

Thomas de la Peña, Carolyn. 1999. "Recharging at the Fordyce: Confronting the Machine and Nature in the Modern Bath." *Technology and Culture* 40(4): 746–69.

Thompson, Edward P. 1978. "Eighteenth-Century English Society: Class Struggle without Class?" *Social History* 3(2): 133–65.

Torres, Nicolás de, and Joséph Dumont. 1762. *Virtudes de las aguas del Peñol, reconocidas y examinadas de orden de la Real Audiencia, por el real Tribunal del Protho-Medicato.* Mexico: Imprenta de la Biblioteca Mexicana.

Tort, José María. 1858. "Sobre la naturaleza de las Aguas de Tehuacán y producción vegetales de sus inmediaciones." In *Suplemento al Tomo Sesto del Boletín de la Sociedad Mexicana de Geografía y Estadística*, 33–41. Mexico: Imprenta de Andrés Boix.

Tortolero, Alejandro. 2000. *El agua y su historia: México y sus desafíos hacia el siglo xxi.* Mexico: Editorial Siglo XXI.

Trawick, P. B. 2003. *The Struggle for Water in Peru: Comedy and Tragedy in the Andean Commons.* Stanford, CA: Stanford University Press.

Tutino, John. 2011. *Making a New World: Founding Capitalism in the Bajío and Spanish North America.* Durham, NC: Duke University Press.

Urbán Martínez, Guadalupe Araceli, and Patricia Elena Aceves Pastrana. 2001. "Leopoldo Río de la Loza en la institucionalización de la química." *Journal of the Mexican Chemical Society* 45(1): 35–39.

Valenza, Janet. 2000. *Taking the Waters in Texas: Springs, Spas, and Fountains of Youth.* Austin: University of Texas Press.

"Valladolid. Análisis de las aguas termales de Cuincho, hecho por la Expedición Botánica de Nueva España . . ." 1790. *Gazeta de México* 4(22) (November 23): 205–9.

Vergara, Francisco. 1892. "Apuntes para la hidrografía médica de Monterrey." In *Estudio sobre la higiene en Monterrey,* edited by Rafael Garza Cantú, 20–33. Monterrey: Ponencia Asociación Americana de Higienistas.

Vigarello, Georges. 1988. *Concepts of Cleanliness: Changing Attitudes in France since the Middle Ages.* Cambridge/Paris: Cambridge University Press / Editions de la Maison des Sciences de L'Homme.

Villaseñor, Federico F. 1900. "Algunas consideraciones acerca del análisis de las aguas potables." *Anales del Instituto Médico Nacional* 4 (August 31).

Virkki, Niilo. 1962. "Comentarios sobre el baño de vapor entre los indígenas de Guatemala." *Guatemala Indígena* 2(2): 71–85.

Vovides, A. P., E. Linares, and R. Bye. 2010. *Jardines botánicos de México: Historia y perspectivas.* Xalapa, Veracruz: Secretaría de Estado de Veracruz.

Wagner, John Richard, ed. 2013. *The Social Life of Water.* New York / Oxford: Berghahn.

Walsh, Casey. 2008. *Building the Borderlands: A Transnational History of Irrigated Cotton along the Mexico-U.S. Border.* College Station: Texas A&M University Press.

———. 2011. "Managing Urban Water Demand in Neoliberal Northern Mexico." *Human Organization* 70(1): 54–62.

———. 2012. "Mexican Water Studies in the Mexico-U.S. Borderlands." *Journal of Political Ecology* 19: 50–56.

Walton, John K. 2012. "Health, Sociability, Politics and Culture: Spas in History, Spas and History: An Overview." *Journal of Tourism History* 4(1): 1–14.

Ward, Evan. 2003. *Border Oasis: Water and the Political Ecology of the Colorado River Delta, 1940–1975*. Tucson: University of Arizona Press.

Weiss, Harry, and Howard Kemble. 1967. *The Great American Water-Cure Craze: A History of Hydrotherapy in the United States*. Trenton, NJ: Past Times Press.

Wilk, Richard. 2006. "Bottled Water: The Pure Commodity in the Age of Branding." *Journal of Consumer Culture* 6(3): 303–25.

Williams, Marilyn Thornton. 1992. *Washing "The Great Unwashed": Public Baths in Urban America, 1840–1920*. Columbus: Ohio State University Press.

Williams, Raymond. 1977. *Marxism and Literature*. Oxford: Oxford Paperbacks.

Wiltse, Jeff. 2007. *Contested Waters: A Social History of Swimming Pools in America*. Chapel Hill: University of North Carolina Press.

Wittfogel, Karl August. 1957. *Oriental Despotism: A Comparative Study of Total Power*. New Haven, CT: Yale University Press.

Wolf, Eric. 1972. "Ownership and Political Ecology." *Anthropological Quarterly* 45(3): 201–5.

——— and Ángel Palerm. 1955. "Irrigation in the Old Acolhua Domain, Mexico." *Southwestern Journal of Anthropology* 11(3): 265–81.

Wolfe, Mikael. 2013. "The Historical Dynamics of Mexico's Groundwater Crisis in La Laguna: Knowledge, Power and Profit, 1920s to 1960s." *Mexican Studies / Estudios Mexicanos* 29(1): 3–35.

———. 2017. *Watering the Revolution: An Environmental and Technical History of Agrarian Reform in Mexico*. Durham, NC: Duke University Press.

Worster, Donald. 1985. *Rivers of Empire: Water, Aridity, and the Growth of the American West*. New York: Pantheon Books.

Zedillo Castillo, A., 1984. *Historia de un hospital: El Hospital Real de Naturales*. Mexico: Instituto Mexicano Del Seguro Social.

Zetland, David. 2009. "The End of Abundance: How Water Bureaucrats Created and Destroyed the Southern California Oasis." *Water Alternatives* 2(3): 350–69.

INDEX

science of mineral waters); sexuality and
nudity views in, 23, 24, 25, 26, 29–31; steam
bathing and *temazcal* in, 11, 17, 23, 25, 28–31,
30*fig.*, 32–33, 38; Tenochtitlán and, 16, 17–20,
21, 32; War of Independence ending, 65
spas: biological and chemical rationale for
development of, 103–12, 104*fig.*, 116–17,
147; industrialization and dispossession of,
124–35; opulence of water for, 67–68, 80–83,
86, 112; tourism based on (*See* spa tourism)
spa tourism: Agua Caliente, 13, 139–40, 140*fig.*,
141; capital for development of, 138–39,
148–50, 161; conflict and compromise over,
138–39, 141, 144–53; cultural and social
constructs influencing, 5, 6, 8, 13, 62–64,
65–66, 140–41, 161; dispossession of waters
and, 124–35, 138–39, 144–48, 153; gambling
and, 141; hydrotherapy in, 147, 156, 157*fig.*;
industrialization and, 124–35; Ixtapan de
la Sal, 13, 138–39, 141–54, 143*fig.*, 155–62;
mercedes of water for, 145–46, 148–53;
municipal bathhouses and, 142, 152*fig.*, 153,
155–56, 156*fig.*, 157*fig.*, 160, 161; Oaxtepec,
153–54; overview of, 13, 137–39, 153–54; public
water infrastructure importance to, 150–51;
science of mineral waters and development
of, 62–64, 65–66, 147; *sonorenses* and, 139–41;
state centralized control of water affecting,
144–46, 147; Tehuacán, 140–41, 158; virtuous
waters for, 155–62
Standing Rock Sioux Tribe Reservation, 9
state centralized control of waters:
industrialization facilitated by, 13, 118–19,
123–24, 129–35; spa tourism affected by,
144–46, 147
steam bathing: cultural and social practices of,
3, 4, 5, 11, 29–31; *hammam* as, 11; immersion
bathing *vs.*, 28–29, 32, 67; opulence of water
and decline of, 67; policing of baths and, 36,
38, 39–41, 42–47, 49, 161; public bathhouses
including, 39–41, 42–47, 49; science of
mineral waters on, 53; Spanish colonial
era, 11, 17, 23, 25, 28–31, 30*fig.*, 32–33, 38;
sweatlodges for, 29; *temazcales* for, 11, 17, 20,
28–31, 30*fig.*, 32–33, 36, 38, 39–41, 42–47, 49,
53, 67, 161; in Tenochtitlán, 20; therapeutic
benefits of, 5, 29, 31, 41, 49
Steward, Julian, 7
Sullivan, Margaret, 76, 77–78
sweatlodges, 29
swimming: bathing relationship to, 17; cultural
and social constructs and popularity of, 5, 42;

licenses for swimming pools, 42; opulence of
water for, 12, 67–68, 72–73, 75, 80, 85, 87–88;
policing of, 42; Spanish colonial era, 24, 26;
spa tourism including, 142, 154; therapeutic
benefits of, 42

Tehuacán: industrialization of waters of, 13,
118, 121–24, 135–36, 137, 159–60; science
of mineral waters of, 12, 59, 101, 102, 112;
Spanish colonial era, 28; spa tourism at,
140–41, 158
temazcales: cultural and social practices of, 11,
29–31; opulence of water and decline of, 67;
policing of baths and, 36, 38, 39–41, 42–47,
49, 161; public bathhouses including, 39–41,
42–47, 49; science of mineral waters on, 53;
Spanish colonial era, 11, 17, 28–31, 30*fig.*,
32–33, 38; sweatlodges for, 29; in Tenochtitlán,
20; therapeutic benefits of, 29, 31, 41, 49
Tenochtitlán, 16, 17–20, 21, 32, 69, 86
tequisquite, 52, 58
Teteo Innan, 29
Tezcatlipoca, 29
therapeutic benefits: bottling mineral waters
for, 13, 118–24, 159–60; cultural practices
and, 3–6, 10; disease concerns *vs.*, 5–6, 25,
48, 71, 91–92, 93, 96–103, 112, 116, 121, 138; of
hydrotherapy, 99–101, 103, 105, 107–9, 112–15,
147, 156; policing water and baths for, 37,
38–39, 40–41, 43, 46, 48–49, 50; science of
mineral waters and, 4–6, 12, 50–66, 91–117,
138 (*See also* biology and water; chemistry
of water); Spanish colonial era views of, 11,
22–25, 29, 31, 38; spas promoting, 103–12,
147–48, 155; spa tourism in relation to, 138,
147–48, 155–62; virtuous waters providing,
10, 51–53, 93, 155–62
Tlaloc, 19–20
Tlatelolco, 68–69
Tocitzin, 29
Topo Chico: industrialization of waters of, 118,
124–36, 125*fig.*, 137, 158, 159–60; science of
mineral waters of, 12, 65, 101, 111
Toro, Ignacio, 83
Torres, Antonio, 83
Treviño, Pedro, 127, 130
Turkish baths, 86, 113

United Kingdom, water and bathing cultures
in, 5, 83
United States: industrialization of waters, 123,
126–27; Mexican-American War with, 137;

www.ingramcontent.com/pod-product-compliance
Lightning Source LLC
Chambersburg PA
CBHW070412270326
41926CB00014B/2789